DATE DUE

JY 27 '95			
AP 27 '98			
JE 9 '09			

DEMCO 38-296

PROFILES IN SUCCESS

Robert E. Bahruth
and
Phillip N. Venditti, Editors

The American Association
of Community and Junior Colleges

Published by The American Association of Community and Junior Colleges
National Center for Higher Education
One Dupont Circle, N.W., Suite 410
Washington, D.C. 20036
Phone: (202) 728-0200

ISBN 0-87117-220-8

Library of Congress Card Number: 90-084156

To the Principles Toward Which

Our Nation and Its Community Colleges

Can Best Dedicate Themselves:

Liberty and Justice for All Humankind

The sight of a tree in full bloom is magnificent. The branches stretch out, like huge arms, reaching for the sky. The leaves, drenched in sunlight, dance in the breeze. Juicy colorful fruit hangs from its branches like Christmas ornaments. However, all its growth and prosperity depends on its roots.

-Shahzad Bhatti
Alumnus of
Kansas College
of Technology

TABLE OF CONTENTS

ACKNOWLEDGEMENTS

By its very nature, this publication owes its existence to the community colleges that graduated its profilees. The presidents and public information officers from the nearly 300 institutions that submitted essays for the collection deserve special thanks for the time and energy they invested in the project. Likewise, the profilees themselves all made special efforts to collect and organize their thoughts carefully prior to writing about their alma maters.

Like most other large-scale writing ventures, this book depended at every step of the way on cooperation and assistance from many people beyond the circle of those who directly wrote or edited parts of it. To the following individuals we extend special thanks and appreciation for such cooperation and assistance:

--John Roueche, director and Sid W. Richardson regents chair; Don Rippey, professor and graduate adviser; George A. Baker III, professor; and Suanne Roueche, senior lecturer; all of the Community College Leadership Program at the University of Texas at Austin, for inspiring understanding and respect for the community college in ourselves and thousands of others around the United States and abroad.

--Stuart Steiner, president, and Don Green, vice president for academic affairs at Genesee Community College, for offering vital editorial advice and overall moral support.

--Ed Gleazer, president emeritus of AACJC—most recently, for writing the thought-provoking preface to this volume; and in general, for reminding the community college movement of the highest ideals it is capable of attaining.

--Dale Parnell, president and chief executive officer, and Phil English, vice president for communications services and technology and director, Community College Satellite Network at AACJC, for supporting the concept underlying the book.

--Bonnie Gardner, assistant vice president for communications, and Susan Reneau, former marketing coordinator, also of AACJC, for lending solidity and coherence to the product through their capable ongoing guidance.

--Alison Anaya, desktop publishing specialist, Ron Stanley, assistant editor, and Michele Jackman, marketing coordinator of AACJC, for putting the physical and word-processed parts of the book into form for publication and developing textual materials to promote it.

--Jim Palmer, formerly vice president for communication services at AACJC and now associate director of the Center for Community College Education at George Mason University, for advocating and gaining preliminary approval for publication of the collection of profiles.

--Mary Lowery, secretary, and Joann Basick, acting director of the Moraine Valley Community College Foundation, for answering questions, as well as soliciting and collecting essays from all over the United States.

--Lee Rasch, formerly senior vice president at Moraine Valley Community College and now president of Western Wisconsin Technical Institute; John Conrath, formerly executive director of resource development at Moraine Valley Community College and now director of educational services at the Management Association of Illinois; Gail Polansky, resource development associate at Moraine Valley Community College; and Jim Young, director of resource development at Pitt Community College, for important early suggestions concerning the dimensions, approach, and format of the book.

--Pat Bauhs, executive dean of community and continuing education at Moraine Valley Community College; Walter Bumphus, vice president of students at Howard Community College; Ron Fracker, grants development coordinator at Washtenaw Community College; "The Block of 1985" from the Community College Leadership Program of the University of Texas at Austin; and Norine Domenico, assistant to the president at the Community College of Aurora, for spreading word throughout the country that profiles of outstanding community college graduates were to be collected for this publication.

--Richard Hart, dean of the College of Education at Boise State University, and Virgil Young, chairman of teacher education, who were helpful in providing creative funding for editorial meetings and postage. They were supportive during the entire course of the project. Jay Fuhriman, director of bilingual education, provided time and moral support to complete this project as well. Also thanks to Susan Stark for helping us to meet the final editorial requirements.

--Carole Simone, Lee Kardis, and Jackie Anderson, all of the Office of Duplicating and Mailing at Moraine Valley Community College; and Nancy Poodry of the Duplicating Services Office at Genesee Community College, for duplicating, sorting, labeling, and mailing materials related to the publication.

--Mike Garrett and Martha Mullen, of Genesee Community College, for designing and producing the cover of the book.

--Bonnie Zalar, secretary to the social science faculty; Karen Barber, clerk-typist in the humanities area; and Melissa Frawley, work-study student at Genesee Community College, for helping to prepare information from the profilees' nomination forms for publication.

--Güisela Bahruth and Yuna Min, the persons with whom we share spousehood, for their patience and willingness to share the hard work associated with putting a book together.

—*Robert Bahruth*
—*Phillip Venditti*

PREFACE

There is no substitute, no matter how eloquent the words, for communication that begins with, "This happened to me," "I was there," "I felt," "I saw," "I know." This book speaks that powerful language of experience. Educational outcomes are revealed in heightened aspirations, enriched lives, and in communities served. "Make the case for community colleges," legislators often request. These personal and poignant statements do exactly that.

I read the essays with particular interest that grows out of many years of interpreting the worth of community colleges to a variety of publics, legislators among them. It was one thing to demonstrate the need for more colleges to accommodate the mounting numbers of "baby boomers" who wanted educational opportunity through the 1960s and early '70s. The community college had appeal to many policy makers because it seemed a less expensive way to meet that need than building four-year colleges. It has been quite another thing to win support for an institution that deliberately seeks to provide educational opportunity to those who are not served by other institutions or who are not served well. Prestige is not often associated with such factors as low cost to the student, open-door admissions, and accessibility to commuters. Frequently in the past (and occasionally now), legislators, citizens in the community, and even some faculty looked forward to the community college's becoming a "regular" college as soon as possible. Often I would wonder when the community college would be recognized as an institution in its own right and with a distinctive educational mission.

The essays in this book reveal institutions that are fulfilling their mission. They are making good on their promise. Results are reported, rather than intentions proclaimed. In many a community where the establishment of a community college has been under discussion, I have been asked to describe the functions such an institution would serve and what would be different about community colleges. This collection of testimonials goes a long way toward the answer to those questions.

A remarkable variety of institutions is represented in the statements of the college alumni. And diversity characterizes the people themselves. However, I find that there are common themes that emerge through all of the dissimilarities. For many reasons these themes deserve our attention. Is it not interesting and important to learn from these students what they considered to be of greatest value in their community college experience? If they were to arrange in rank order their perceptions of the relative value of college contributions toward what they have become, would there be a match with our views of priorities in the college program? Would the student views square with board perceptions or those of the legislators? I wonder, too, if the perceptions of the alumni as registered in their statements are different now from what they would have been at the time they were students. There is material here for a good deal of profitable study and discussion, as well as stimulation to those engaged in institutional research and marketing.

However intrigued we may be with the many ways these essays can be of value, one theme clearly is paramount. There is a common refrain: If the college had not been there, very likely my life would have been greatly diminished. In many ways the authors say, "I would not be where I am today." "It has been a college of hope; without it, where would I be?" "The college was the door to my future."

These words gratify those of us who are committed to community college services. But they are also sobering words, for they raise questions in our minds of what potential would remain untapped, what possibilities stunted both of individuals and the community, if the college had not been there with its receptive and encouraging stance. And that leads us to wonder about other people who could be like those whose stories we have read, but who have not been reached. And what of the future? Surely the times in which we live require more than ever colleges that will offer learning experiences described by these essayists as essential to what they have become.

This book is a good idea. There is freshness to it. The stories are interesting, and for the discerning reader there are many clues to making community colleges even better than they are. We owe thanks to the originators of the idea, to the cooperating institutions, to those who wrote the statements, and to AACJC as publisher.

Edmund J. Gleazer, Jr.
President Emeritus
American Association of Community
and Junior Colleges

CHAPTER ONE

What This Book Is About

We conceived the idea for this book at the conclusion of an especially invigorating professional event: the 1988 International Conference on Teaching Excellence in Austin, Texas. Standing outside the hotel where that gathering had taken place, we reflected on what each of us owed to community colleges.

For one of us, a community college education some 20 years before had helped transform a youth's uncertainty and limited vision into a sense of self-confidence and hope for the future. The essay on page 224 describes this transformation. For the other of us, mid-life exposure to the community college as a teacher and administrator reconfirmed a conviction that equity, community progress, and international fellowship can be approached in this nation. Collecting the essays in this book has been a way for us to show convincingly just how widespread, representative, and important these kinds of experiences have become all over the United States.

We hope to reach several kinds of people with this publication: those who need encouragement to fulfill personal goals; those in decision-making positions who can help influence the fate of community colleges; and those already associated with community colleges who can appreciate and celebrate their successes.

Many readers of this book, we hope, will resemble the hundreds who wrote for it. People whose self-confidence is low and who aren't sure they can succeed in college may be inspired by the fact that most of those who wrote these essays once felt the same way. People who don't know if their career directions are what they want them to be may also be inspired; many of the essayists weren't sure, either. And people who have known nothing about the diversity and receptiveness of community colleges may be inspired by this book; every one of the essayists was once just as unaware. In short, this book should mean something to ordinary men and women who want to attain their full potential and can do so if they are offered the right opportunities.

At the same time, we believe that a publication of this sort will attract and reinforce the support of community members for community colleges in general and for the colleges represented here in particular. To realize its promise, this book needs to reach state legislators, mayors, and other local and state officials; if it does, we're convinced the long-term results will be favorable. Likewise, we hope the stories told in these pages will strike responsive chords within high school counselors and four-year college instructors and administrators. These groups and others like them, because they compose the world in which community colleges exist and upon which they depend for sustenance, are logical audiences for this book.

DEVELOPMENT OF THE IDEA

With support from the American Association of Community and Junior Colleges, our idea in Texas evolved from a dream into a reality. In September 1989, we sent letters to the presidents and public information officers of each AACJC-member college. We asked them to identify outstanding graduates of their colleges who might be willing to compose brief essays discussing what the community college experience meant to their development.

Within 10 weeks, more than 190 colleges in 39 states answered our letter. With their responses, the colleges forwarded essays by nearly 500 of their graduates. We then set about reviewing and assessing this pool of essays, knowing that it would be necessary to winnow it down to the number eventually to be published.

According honors of any sort is generally difficult because so many good candidates compete for them. As we selected essays for *Profiles in Success*, this truism assumed special intensity. The writings we received from around the country, almost without exception, exemplified high standards of excellence and impact. All we needed to attend to as editors, aside from the overwhelming formidability of choosing only one of every two essays submitted, were such minor matters as consistency in length and writing conventions from profile to profile. The selection criteria we employed included writing quality, broad geographic distribution, and representativeness of the different populations served by community colleges.

Because they serve local communities, two-year institutions make it possible for older, working students to accommodate their courses of study to the demands of their daily lives. In selecting the essays, we found it relatively easy to reflect the presence of such students among the graduates of participating community colleges.

In our opinion, each of the essays we received in 1989 testifies directly and convincingly to the message we had hoped would emerge from the collection as a whole. Every essay has the potential to touch and alter the lives of those who read it, since it reflects the experience of a person who made personal strides as well as important contributions to society. And each of the essays may be used effectively by its author's alma mater as a tool for showing prospective students what a community college education can do for them.

A special note of explanation should be sounded with regard to the profilees in this book. Many of these individuals are widely known and respected within their professional fields. In fact, the names of a few approach being household words in some parts of the country. Nevertheless, we neither solicited nor selected profilees on the basis of how well-known or famous they might be. Instead, our aim was to illustrate that society benefits in a multitude of ways—some obvious and broadly noted, some more private and unsung—from the special talents of its community college graduates.

ORGANIZATION OF THE MATERIAL

The essays composing this publication have been arranged into eight chapters representing categories of careers which the essayists have entered: law; science, technology, and agriculture; entertainment, media, and sports; health and medicine; social services; business; public service; and education. A photograph, along with a brief statement describing the profilee's biographical information and contributions to society, accompanies each essay.

Colleges—large and small, urban and rural, new and old, financially comfortable and struggling—all are represented here. Readers who'd like to locate profiles from particular states may consult the college index that starts on page 301.

HINT FOR THE READER

This book is something like a treasure chest. It contains hundreds of sparkling gems, each of which is splendid and complete in its own way. But like a chest brimming with gems, the essays here are probably examined and appreciated better in handfuls than all at once. We suggest that readers sample the essays in clusters, perhaps jumping from chapter to chapter, since doing so will provide a flavor of the diversity represented by the writers and their experiences.

CHAPTER TWO

Community Colleges:

An Introduction for the Uninitiated

In these times of national and international challenge, community colleges are becoming increasingly more significant to American society. Our conviction is that no more powerful and persuasive arguments can be found in support of community colleges than the words of their own graduates, and that is why the rest of this book is made up of essays by some of the most impressive of those graduates.

Nevertheless, knowing something of the broad background and context of American community colleges—some statistics, some history—can further substantiate their value. Likewise, responding to some prevalent concerns about what community colleges do is an important part of establishing their worth.

WHAT MOST PEOPLE DON'T KNOW ABOUT COMMUNITY COLLEGES

Despite the abundance of community colleges and their accessibility to nearly every citizen in the country, various factors combine to make them less known and understood than most other American educational institutions. Among these factors are the relative newness of community colleges, the complexity of their goals and missions, their constant evolution, and the fact that they serve diffuse and diverse populations.

History of Community Colleges

As early as the mid-nineteenth century, about the time that the Morrill Act created land-grant institutions as "the people's colleges" in 1862, several presidents of major American universities were calling for the establishment of a two-year college. The major purpose of this new kind of educational institution would be to prepare students for the transition from high school to university by offering all of the lower-division courses usually taught during the freshman and sophomore years. Thus, two-year colleges would free universities to focus exclusively on upper-division studies and the rigors of research. William Rainey Harper, president of the University of Chicago, attempted to implement this philosophy by founding the first junior college at his institution in 1896. Five years later, Joliet Junior College opened its doors as the first public two-year college in the United States.

As is well known, however, few universities in years since have yielded their reign over lower-division studies. The institutionalization of the two-year college as an independent link in the educational chain, acting as the sole educational resource for students beginning postsecondary education, did not occur. Instead, two-year colleges came to be viewed in many quarters as adjunct institutions responsible for educating poorly prepared students, providing vocational training and continuing educational opportunities. To combat this image of poor quality, and to seek a more equal status with universities, many two-year schools have sought over the decades to transform themselves into four-year institutions.

Despite those early difficulties in earning credibility from other segments of postsecondary education, two-year colleges have come to provide services upon which we all depend and from which we all benefit. Indeed, they have become institutionalized by popular demand.

A major influence on the positive development of two-year colleges came during the administration of Harry S. Truman. In 1947 the Truman Commission on Higher Education had the foresight and vision to recognize the potential of these institutions to provide access to postsecondary education for previously unrepresented portions of the American population. Among other things, the commission recommended changing the name of two-year colleges from junior to community college in order to more accurately reflect the role they would increasingly come to fulfill.

In addition to the community college's promise of open access, growth has been a hallmark of the institution since its inception at the beginning of this century. A total of only nine junior colleges existed in 1909, whereas today more than 1,200 community colleges dot the map in every state but one. Enrollments in community colleges grew by 215 percent between 1965 and 1975, the "boom" years during which a new college was opened on an average of once a week. In recent years, the number of students attending community colleges has continued to climb at a steady, though slower, pace.

Similarly, the number of associate degrees awarded by community colleges grew by more than 24 percent between 1975 and 1985, from 360,171 to 446,047 (Palmer, 1987-88). To put this expansion in perspective, the number of bachelor's degrees awarded increased by only 7 percent during the same period, and the number of graduate degrees actually declined.

Community colleges have become, in the less than one century in which they have existed, a potent new social force. Unique among all segments of postsecondary education, community colleges can—and do—consider themselves part of a "movement."

Characteristics of Community College Services and Enrollments

Almost all community colleges share certain characteristics that set them apart from other educational institutions. Following is a brief list of such distinguishing characteristics:

- All together, community colleges educate an enormous number of people. By the end of the 1980s they enrolled between five and six million individuals in credit courses each year, as well as somewhere over four million others taking non-credit classes. Those enrolled in credit courses in community colleges represent 41 percent of the entire enrollment in all undergraduate programs across America, and community colleges are the institutions of choice for over 55 percent of all first-time postsecondary students.

- Tuition for students who attend community colleges is lower than that charged by other postsecondary institutions, including state-supported four-year colleges and universities. The overwhelming bulk of community college students commute to and from school, which affords them a further substantial financial advantage over enrolling at residential campuses.

- Diversity in the size of America's community colleges is apt to strike even the casual observer. Almost half of the more than 1,200 public and private community, junior, and technical colleges serve rural areas not accommodated by other postsecondary institutions. In effect, these schools have been brought to people who might not otherwise have been able to leave home to participate in postsecondary education. Rural colleges like these may serve as few as 100 students; in fact, one quarter of all community colleges enroll fewer than 1,400 students. In contrast, many of the larger community colleges serve urban areas. Among the largest multi-college and multi-campus systems—the 25 percent of public colleges whose enrollments exceed 5,750 students—are the Los Angeles Community College District, with around 100,000 students; the Maricopa Community College District in Arizona; the City Colleges of Chicago; the Dallas County Community College District; and Miami-Dade Community College in Florida.

- Larger proportions of older students are enrolled by community colleges than by higher education as a whole. The average age of students in credit classes at public community colleges is 28, and 50 percent of all community college students are older than the traditional 18-to-24-year-old college cohort. Day classes reflect an average age of 23, whereas evening classes reflect an average age of 38.

- Community colleges serve large numbers of minority students. In fact, they have been referred to as "the Ellis Island of higher education" (Vaughan, 1983, p. 9). By fall 1986, minorities composed approximately 22 percent of all students in community, technical, and junior colleges. More than 57 percent of Native American college students, 55 percent of Hispanic college students, 43 percent of all Black college students, and 41 percent of all Asian college students attend community colleges (National Center for Higher Education, 1989). When compared to four-year college and university enrollments, this represents a disproportionate share of minorities served.

- The proportion of handicapped students is much larger in community colleges than in four-year colleges and universities. International students make up a large portion of students attending two-year schools: nearly 1.1 percent (Cohen & Brawer, 1989).

CONCERNS ABOUT COMMUNITY COLLEGES

As the number of community colleges has grown over the years, and as the magnitude of support required to operate them effectively has expanded, concerns have been raised about their nature and worth. These concerns deserve examination.

"The quality of community college students is low." Some people have referred to community colleges as "high schools with ashtrays." Terminology like this, despite its disparaging tone, does reflect the fact that community colleges share with public education a willingness to accept the least-prepared students along with the most proficient ones.

No knowledgeable person would dispute the statement that many community college students are less academically able than are their four-year college counterparts. The fact that community colleges accept students who are excluded elsewhere because of their academic deficiencies means that virtually all such colleges provide developmental courses intended to help "at-risk" individuals prepare for educational success. In 1982, 25 percent of entering full-time community college students said they had ranked in the top fifth of their high school classes; 60 percent of those entering universities made the same claim (Astin, Hemond, & Richardson,1982).

These statistics apply only to students entering college for the first time, however—meaning less than a third of the community college population. Two-thirds of community college students, including many who have already received baccalaureate degrees, possess academic abilities and create influences on the educational climate of their colleges which have yet to be assessed (Deegan & Tillery, 1985). In fact, one feature which attracts both students and staff to community colleges is that their clientele ranges broadly enough in age and academic readiness to make almost anyone feel at home.

Furthermore, community colleges rightly ask to be judged by the quality of their products, not by nature of the "raw material" from which those products derive. If they are assessed in terms of "value added"—that is, what they are able to do to change the lives of the students they enroll—community colleges can take pride in the kinds of personal successes typified by the profilees in this book.

At the same time, it is fair to say that community colleges have plenty of work to do. Part of their challenge is to make certain that the nature of educational opportunity for their students transcends "the right to meet minimal standards" (Cross, 1976, p. 3) and constitutes, instead, the right to develop one's talents to the maximum possible extent.

"Transfer rates are low." Accurate postsecondary transfer rates are hard to come by in this country because data are not regularly collected by either government agencies or four-year colleges and universities. Furthermore, definitions and methods of calculating transfer rates differ substantially from source to source.

Data from one major study (Brown, 1987) did indicate that approximately 29 percent of 1980 high school graduates who entered two-year colleges by October 1983 later transferred to four-year institutions. One reality which these figures fail to reflect, however, is that many students who enroll at community colleges never intend to transfer. In fact, studies in at least five states throughout the 1980s revealed that only about one third of all entering community college students referred to earning a baccalaureate degree as their primary aspiration (Cohen & Brawer, 1989).

The objectives that some community college students pursue may be attained through taking courses whose completion leads directly into the work force. Earning an associate in applied science degree in some technical field is a goal for many of these individuals, and certificates requiring less than a full two years of study are also sought by many.

Large numbers of other students, knowing that community colleges will accept them whenever they enter and allow them to "stop out" later without penalty, may deliberately sample a handful of courses over a space of several years without ever planning to earn a diploma. Perhaps as much as a quarter of those who attend community colleges may fall into this "special interest" category; that is, their aims are to enlarge their range of experiences, improve their language skills in English or some other language, or satisfy a desire for some sort of personal growth (Sheldon,1983). Men and women like these often continue to derive benefits for decades from associating with their community colleges, regardless of whether they ever transfer elsewhere.

Another point to consider in assessing transfer and graduation statistics in community colleges is that the vast majority of bachelor's degree recipients take more than four years to graduate. Of 1980 high school graduates who enrolled full-time in a "four-year college" during the fall of 1980, for instance, only about one in five had earned a baccalaureate degree nine semesters later (Palmer, 1990). In short, the trend among America's college students— wherever they go to school—is to depart freely and frequently from the constrained schedules of the past.

"Community colleges perpetuate class and economic stratification by `cooling out' students." Statistics indicate that many community college students have less-educated parents and come from families with lower incomes than do students who attend four-year colleges and universities. In a study of the nation's high school graduating class one year, only 27 percent of students who entered community colleges were from families in the top quartile on a social class index, whereas 61 percent of those who entered four-year institutions came from that group (National Center for Education Statistics, 1984).

Such differences in socioeconomic status have led some critics to conclude that a prominent underlying function of community colleges is to channel students into their "appropriate" positions in the social order, based upon their past status. Zwerling (1976), for instance, asserted that students who experience community college education are apt to reach for less than they are capable of accomplishing by, for example, being counseled into "lower-status" terminal occupational curricula rather than baccalaureate transfer programs.

What critical analyses like these overlook is that the choice for the vast majority of first-time community college students is not between the status-enhancing possibilities associated with attending a four-year college or university on the one hand, and the supposedly less lofty outcomes of a community college education on the other. Instead, the choice is generally between the community college and nothing. Were it not for the community college, these millions of men and women might never even realize that they have opportunities for personal and vocational growth, much less find ways to explore such opportunities.

Furthermore, one of the central strengths of the community college as a tool for personal improvement is the variety of curricula it offers to its students at any stage in their lives. The 20-year-old man or woman who is "cooled out" enough upon leaving a community college to settle into a technician's position in some business or industrial firm, for example, will have many chances later in life to upgrade job skills or embark upon an entirely different course of study—perhaps leading to a four-year degree at that time—by returning to the community college at will. In other words, the "allocative function" of community colleges (Cohen & Brawer, 1989, p. 355) is frequently able to expand and enrich people's life chances. In essence, the community college is an extraordinary enabling institution.

CONCLUSION

After reviewing the essays presented in the remainder of this volume, readers will readily appreciate that community colleges provide opportunities for people to grow. Comments by the profilees exemplify how students with resolve can and do embark upon careers after graduation from two-year colleges—and then go on to make significant contributions to their communities and our nation.

These profiles demonstrate that community colleges most often offer a hand up, not a hand-out. They show how community colleges assist people to reach the goals they are willing to work for—without compromising standards. They reflect the myriad ways in which we all benefit from the services provided by community colleges. And they show that, were it not for community colleges, many of the dreams of our nation's most responsible and productive citizens might never have been conceived, much less brought into reality.

REFERENCES

Astin, A., Hemond, M., & Richardson, G. (1982). *The American freshman: National norms for fall 1982*. Los Angeles: University of California.

Brown, S.V. (1987). *Minorities in the graduate education pipeline*. Princeton, NJ: Educational Testing Service.

Cohen, A, & Brawer, F. (1989). *The American community college* (2nd edition). San Francisco: Jossey-Bass.

Cross, K. (1976). *Accent on learning*. San Francisco: Jossey-Bass.

Deegan, W., Tillery, D., & associates (1985). *Renewing the American community college: Priorities and strategies for effective leadership*. San Francisco: Jossey-Bass.

National Center for Education Statistics (1984). *Two years after high school: A capsule description of 1980 seniors*. Washington, DC: National Center for Education Statistics.

National Center for Higher Education (1989). *A summary of selected national data pertaining to community, technical, and junior colleges*. Washington, DC.

Palmer, J. (1990, March 27). "The 'four-year' degree and other higher education myths." *The Community, Technical, and Junior College Times*, p. 2.

Roueche, J. & Baker, G. (1987). *Access and excellence: The open-door college*. Washington, DC: Community College Press.

Sheldon, M. (1983). *Statewide longitudinal study*. (Part 5). Los Angeles: Los Angeles Pierce College.

Vaughan, G. & associates (1983). *Issues for community college leaders in a new era.* San Francisco: Jossey-Bass.

Zwerling, L. (1976). *Second best: The crisis of the community college.* New York: McGraw-Hill.

CHAPTER THREE

Law

Connie Bushnell
El Centro College, TX

In 1980, I was a legal secretary preparing documents for real estate transactions in a branch office of a title company in Dallas, TX. My 16 years' experience in real estate law did not give me the credibility I needed to go forward with my career in law. It was evident that I would need to return to school for my associate's degree in the legal assistant profession.

College was not new to me, since I had completed one year of college at North Texas University in Denton and one year of courses at El Centro Community College in Dallas. I was relieved to see others, my age and older, who had also found doors closed because of a lack of education. I had never been a top student in high school or college. After completion of my first semester at El Centro in the legal assistant program, I was surprised to discover that I really could make A's in all my subjects. It spurred me on to learn all I could in my chosen profession. I knew El Centro was the right school, and I had chosen the right profession.

Upon graduation in 1983, I was hired as a paralegal in the law firm of Akin, Gump, Strauss, Hauer & Feld. I passed the National Association of Legal Assistants' exam for national certification and received my Certified Legal Assistant Certificate in 1988. It is the only national certification examination given to legal assistants.

Recently, my firm asked me to chair a committee to establish a continuing legal education program for legal assistants in the firm within six sections covering five substantive topics of law. In addition, I was invited to serve during the academic year of 1989-90 on El Centro College's Legal Assistant Advisory Committee.

The encouragement from my family gave me courage to go back to school and to continue my studies. The help I received from my teachers gave me the inspiration to continue in the program and receive my associate's degree in applied arts and science. I felt challenged and stimulated intellectually. Upon graduation, I felt prepared to succeed. Also, I felt prepared and inspired to continue with my education to reach my goal of someday being a practicing attorney in the city of Dallas. Presently, I am attending the University of Texas at Dallas for a bachelor's degree in general studies.

Connie Bushnell graduated from El Centro College in 1983 with a degree in legal assistantship and is now living in Dallas, TX. As the leader in the legal assistant department of Akin, Gump, Strauss, Hauer & Feld--the largest law firm in Dallas--she conducts a continuing legal education program for legal assistants through the Dallas Bar Association. She also serves on the El Centro Legal Assistant Advisory Committee, providing valuable insights for the college's instructors and students.

Going to school at night is difficult, but I am dedicated to attaining my goals and to helping others to become successful legal assistants. I hope I can pass on to others in the profession or entering the profession the encouragement and help that I was given at El Centro. I will always be thankful.

James M. Graves, Jr.
Muskegon Community College, MI

Muskegon Community College has had a more significant impact on my life than any other institution. Just as for many of the blue-collar young people of our industrial town, college seemed an impossibility to me. The encouragement that so many of our elders and teachers gave us to stay away from the vagaries of the assembly line and grow to our fullest intellectual and social potential seemed eroded by the cold economic realities of high college costs and modest working class wages. "JC," as the local community college was known, had been available for several years with its rock-solid reputation in both college transfer and vocational programs. Nevertheless, I felt that if I was going to exert what seemed at the time to be a massive effort to obtain a college education, I wanted the glamour of going away to a four-year college in another state. The sages noted, "A prophet has no honor in his own country," and many young people possess that same attitude toward their home communities and local educational institutions. I certainly did.

A combination of scholarships, summer jobs, and a work-study program enabled me to enroll in an excellent four-year school. However, a college located in a rural setting with a student body smaller than my high school graduating class was not satisfactory to me, and a refrain familiar to many freshmen appeared: College for what? What career, what goal? Escape from the humdrum of 40 years' operating a punch press was a powerful initial motivator, but escape to what? The confusion was so great that in spite of the encouragement of kind professors and the satisfaction of adequate grades, I left after one semester to work, save money, and sort things out.

Neither my home community nor the local "JC" seemed so boring now. Indeed, they seemed downright attractive. Encouraged by the wise counsel of former teachers, I enrolled as a full-time transfer student at Muskegon Community College and continued my part-time job. It was one of the most fortunate decisions I ever made. The basic two-year college transfer program enabled me to progress toward my degree while engaged in my "sorting out" process for career goals in the familiarity of my own community. I can honestly say that the quality of teaching at the community college was better than that of the prestigious four-year college, graduate school, and professional school that I eventually attended. The faculty contact, faculty encouragement, and academic standards were second to none.

The study skills, work habits, and motivation for self-help instilled by the college have lasted a lifetime.

Thanks, JC, for being there!

James Graves, Jr., graduated from Muskegon Community College in 1962 with a degree in liberal arts. He lives in Muskegon, MI, where he serves as judge of the 14th Circuit Court. In this capacity, he has often presented crime prevention programs at area schools. He also serves on a Law Day panel each year and has contributed substantially to the activities of the Child Abuse Council and various local churches.

Russell J. Ippolito
Westchester Community College, NY

They say that community colleges perform a "salvage" function. It's true: I was "salvaged" at Westchester Community College, NY. By the fourth grade, I had failed every class. By the time I was a senior in high school, I had read one book. Sitting second to last in my class, I dropped out my senior year.

After dropping out, I worked many different jobs. I guess I became what people call "street wise." Although there is a great deal to learn in the street and old men can impart wisdom, something in my life was missing. I decided to go to college.

With a great deal of effort, I was able to get a high school equivalency diploma. I applied to several colleges and was turned down summarily. Several years later I heard that Westchester Community College had open enrollment. I applied, took a course, and a passion of knowledge flared. I enrolled full-time and through several twists of fate became the student government president.

I graduated from Westchester Community College with high honors, transferred to Columbia University, and graduated with high honors in philosophy. I entered Pace University School of Law, and currently sit on the Ranking Scholars list.

The encouragement I received at Westchester Community College was inspiring. I was not sure I could handle college, but my professors assured me I could. Their guidance and support still drive me. I have not taken a course or read a book totally believing I could finish. Their voices echo in my mind telling me I can.

I can now try to help society by litigating in environmental law or trying to light a student's mind ablaze by teaching. Whatever I decide, there is one thing for certain, these choices were not mine to make before I found Westchester Community College.

Russell Ippolito graduated from Westchester Community College in 1985 with a degree in liberal arts/social science. He now lives in Tarrytown, NY. After serving as president of the Westchester student senate for two years and graduating with a 3.86 GPA, he earned a BA degree in philosophy from Columbia University and is now a second-year law student at Pace University. He has attained all these achievements despite a learning disability.

Wesley Johnson
Central Community College, NE

I can summarize how my junior college experience contributed to my life in a single word: Opportunity—opportunity to discover myself, opportunity to discover a rewarding career, and the opportunity to contribute to the people I represent as an attorney.

I come from a working, middle-class family, and was a mediocre high school student. I had few, if any, goals other than graduation as soon as possible. I graduated at mid-term of my senior year and quickly realized I wanted and needed more formal education.

My community's junior college, Central Community College - Platte Campus, was the answer to my needs. Platte provided formal, post-high school education, was close to home, had reasonable academic admission requirements and affordable tuition rates, and, quite unexpectedly, presented me with the opportunity to discover myself.

With the guidance and assistance of concerned and dedicated Platte instructors, I found that I like to think, to analyze, to organize my thoughts in some coherent manner. In short, I discovered how much I really wanted to learn. Consequently, I attained the scholastic prerequisites necessary for admission into a four-year university and, later, law school.

At the university and especially the law school I attended, my peers were graduates of educational institutions more established and renowned than Platte. Nevertheless, my junior college foundation served me well. Academically, I was competitive, and I discovered the opportunity for a rewarding career in law.

As an attorney, my practice is substantially limited to personal injury and workers' compensation claims. All my clients suffer from injuries, many so seriously injured they cannot continue to be employed in their present occupations. They want and need the education and retraining necessary to acquire new employment skills tailored to fit their post-injury capabilities.

My junior college experience provided me a unique, personal opportunity to sincerely urge these clients to attend local junior colleges for their educational and vocational needs. I can honestly discuss with these individuals what they can expect and the rewards they can achieve through a junior college experience like my own.

My life is personally, professionally, and socially more satisfying and productive because of my junior college experience and the opportunities it provided and continues to provide to me.

Wesley Johnson graduated from Central Community College--Platte Campus in 1975 with a degree in social science. He now lives in Dallas, TX. After distinguishing himself as an undergraduate student, he spent several years in rural Texas as a VISTA volunteer. As a practicing attorney in Dallas, he continues to exhibit a strong social conscience and has recently defended several cases dealing with environmental protection issues.

Mary Kegley
Wytheville Community College, VA

Wytheville Community College provided me with an opportunity that I had missed when I graduated from high school in Canada. After one year of college there I became a medical laboratory technician, as there was no money for further college work. Over a game of bridge in the 1960s, I heard a girlfriend say that she had gone back to school at WCC. That was amazing to me, as she had several children and was probably in her 30s. Then I realized if she could do it, maybe I could too, even though I had two children. After an hour conference with the president of the college, and a great deal of encouragement, I was sure I could do it.

As I was interested in history, my first class was American history. The next year I added other courses, and before I knew it I was going full-time. It was an exhilarating time. After graduation in 1969, I completed one quarter at Virginia Polytechnic Institute and State University but had to postpone my studies until 1974, when I resumed classes at Radford University, VA, remaining there to take my graduate work in history and obtaining my master's degree in 1975.

In 1978, I took another step--entering law school in Richmond, where I graduated with a juris doctor degree in 1981. A few months after graduation, I began teaching business law at Virginia Tech, where I am still employed in the department of finance. I am also in the general practice of law and have my own office in Wytheville. My interest in history has continued, and in 1989 I completed *Wythe County, Virginia, A Bicentennial History*, my eighteenth history-related publication. Without WCC, none of this would have been possible. I will always be thankful for the comments made over the bridge table, the encouragement I received from the president, and for the presence of WCC in the community.

Mary Kegley graduated from Wytheville Community College in 1969 with a degree in liberal arts. She now lives in Wytheville, VA, where she has written several books about that area's history and genealogy. In her role as an attorney, she has taken an active part in many community organizations. Throughout her life, she has contributed to society while successfully handling the responsibilities of a home and family.

Maurice Ulric Killion
Kaskaskia College, IL

Upon graduation from Centralia High School, IL, in 1971, I did not immediately opt for a college education. I found employment with the Illinois Department of Transportation as a civil engineering technician trainee. One year later, however, I did enroll as a full-time student at Kaskaskia College, IL.

I recall the student body as being friendly, and the instructors and staff as being understanding and encouraging. The teaching methodologies were diverse, and I did receive a strong liberal arts education. Upon graduation from Kaskaskia College with an associate of arts degree, I found myself well-prepared for life and the pursuit of higher education.

I went on to enroll at Illinois State University. As an undergraduate at ISU, I was awarded an undergraduate teaching assistantship in the department of economics and was subsequently awarded a bachelor of science degree in economics.

Upon graduation from ISU, I was selected as a Counsel on Legal Educational Opportunity student (CLEO) and went on to successfully complete the CLEO program conducted at the University of Cincinnati School of Law, OH. In 1979 I was awarded a juris doctor degree from the Capital University Law School in Columbus, OH.

In 1987 I returned to the Kaskaskia College District. As a member of the bar of the state of Illinois, I have served this community in several capacities. I was an instructor in the Administration of Justice Department at Kaskaskia College, and I am presently serving on the college's Administration of Justice Advisory Committee. I have been professionally employed by both the Clinton County and Marion County State's Attorneys Office. Upon my resignation as assistant state's attorney for Marion County in November of 1989, I was appointed to the position of Clinton County public defender.

It is simple to list my accomplishments. It is difficult, however, to articulate my appreciation for Kaskaskia College. As a young student, I did undergo an experience at Kaskaskia College that left me confident in the pursuit of higher goals. Thanks to the Kaskaskia College District, I returned and continue to serve this community in my professional capacity as a lawyer.

> *Maurice Ulric Killion graduated from Kaskaskia College in 1974 with an AA degree. He now lives in Centralia, IL, where he has served as assistant state's attorney for Marion County and as Clinton County public defender. He has taken part in the activities of his alma mater over the years as a member of its Administration of Justice Advisory Committee and as an instructor in its Administration of Justice Department.*

Renae Kimble
Niagara County Community College, NY

Within a few days of my arrival at Niagara County Community College a meeting was called to orient freshman minority students to college life and to introduce the Black Student Union. An election was held for officers, and I was surprisingly nominated for vice president. I did not stand a chance in the world of winning, because my two opponents were sophomores. To my surprise, however, I won the election and have been thrust into leadership positions ever since.

NCCC allowed me to freely express myself and to discover and develop leadership, public speaking, and organizational skills that were virtually untapped resources within me.

In order to meet the academic demands of any institution of higher learning, one needs to be disciplined, self-motivated, confident, and have mental toughness. NCCC's faculty helped to provide a balanced educational program for the students. This program provided enough challenges in the classroom environment to allow anyone to develop intellectually as well as develop written and oral skills necessary to succeed in one's own community and have a rewarding career.

I graduated from SUNY at Buffalo with a bachelor of arts degree in political science and then graduated from SUNY at Buffalo Faculty of Law and Jurisprudence with a juris doctor degree. I am employed at the Niagara County Attorney's Office as a legal associate.

I have taken the NCCC commitment to excellence and applied it to every phase of my life as a community leader and activist. I have been privileged to speak at rallies of thousands, including a rally for Jesse Jackson and a rally for the official recognition of Dr. Martin Luther King, Jr.'s birthday.

I have been the first Black woman to run for city council and have organized and spearheaded the fight to change the present at-large system of electing city council candidates to district council representation, thereby eliminating the disenfranchisement experienced by residents who live outside of the most affluent section of town where the majority of the council representatives live. I have spoken to young people about saying "no" to drugs. And I have worked to develop a magnet school program. I hope to give back to my community just a little bit of what NCCC has given to me.

Renae Kimble graduated from Niagara County Community College in 1974 with a degree in liberal arts. She now lives in Niagara Falls, NY. The numerous community boards on which she has served include the Red Cross, the Niagara Community Center and Girls Club, the Black Women's Political Network, the Renaissance Women of Buffalo, the Niagara Christian Fellowship, and the New York State Democratic Committee.

I will never forget my educational experience at NCCC. One may receive numerous degrees, awards, and commendations in a lifetime, but one must always look back to the place where it started. For me, it was NCCC. Thanks, NCCC, for providing me a golden opportunity to pursue and achieve academic excellence and a more fully empowered life.

Kenneth H. Laudenslager
Kalamazoo Valley Community College, MI

After quitting high school in the 10th grade, I enlisted in the U.S. Army on my 17th birthday. I had hoped the military background would provide some hope for my seemingly rocky future. Upon returning from Vietnam as a decorated soldier I quickly learned my future in society was little, if at all, affected by the wounds and medals I received overseas. I needed a vehicle to success.

Despite coming from a family of honorable factory workers with no college experience, I felt directed by my family to pursue college. Kalamazoo Valley Community College had just completed its first year of instruction, and since it received rave reviews from our community, I explored enrollment.

My fears of failure, along with residual problematic emotions stemming from Vietnam, quickly evaporated when I first entered KVCC's Redwood Hall. The numbing effects of self-doubt and the war experience dramatically eased with the tide of personal encouragement I received from KVCC's staff and administration. The counselors met my ambition with a well-balanced curriculum at an affordable price. I applied a determined discipline to my studies, graduating two years later with a 4.0 grade point average.

KVCC also was the springboard for my social adjustment. The school's positive approach in dealing with the students and their needs prompted me to run for president of the student body. After being elected, I sat with the board of trustees and offered student responses to the board's mapping of this institution's future.

Thanks to my academic achievements at KVCC, I was given a fully paid scholarship to Western Michigan University. I graduated there in 1973 summa cum laude. With KVCC's motivational momentum, I applied for and received a scholarship at the Ohio Northern University College of Law. Applying the organizational skills and confidence cultivated at KVCC, I became vice president and president of the student bar association and was named research assistant to the dean.

I graduated cum laude from Ohio Northern University in 1976. I entered the corporate world, and by 1978 I was president of WD Design Corporation in Alhambra, CA, vice-president of marketing and sales for Kazoo Imagination Toys, and regional manager for the parent company, Merchants Publishing Company of Kalamazoo. Since 1978 I have engaged in the private practice of law and currently own a successful trial law firm. The communication skills and confidence acquired at KVCC are seen in action every day of my life, socially or in the business setting. Today, when I direct clients to KVCC, I feel a warm glow inside from the ember of success that that institution once kindled in me. Thanks Valley!!!

Kenneth H. Laudenslager graduated from Kalamazoo Valley Community College in 1971 with a degree in community service/psychology and is now living in Kalamazoo, MI. He is a member of the Michigan Trial Lawyers Association and the Kalamazoo County Trial Attorneys' Association. Kenneth also performs numerous public service activities, such as visiting high schools and speaking to students on Career Day. He now holds the position of trial attorney.

Audrey L. Lloyd
Eastfield College, TX

Eastfield College is much more than a place to learn academics. It provides an environment where qualified instructors, counselors, and other education professionals encourage people of all ages who have reached a point in their lives at which they need nurturing, guidance, information, and good advice in order to create their future.

At Eastfield I received a good education in basic academic subjects, which I was able to add to and build upon from that point on. The instructor-to-student ratio was low enough to enable instructors to spend time with students, a big benefit for a young person who is learning to make decisions that will affect the rest of her life. One particular instructor who is also a counselor at Eastfield, A.W. Massey, spent a great deal of time helping me to find some direction for my life. I benefitted greatly from the encouragement and information he gave me.

Perhaps because Eastfield is a small school, it is easy to become involved in on-campus activities, an important way to learn leadership and responsibility. My experiences as a member of Phi Theta Kappa demonstrated to me that if I worked hard there would be rewards. As a result of my having been a member at Eastfield, I qualified to be a member of Phi Theta Kappa Alumni at Southwest Texas State University. I was fortunate enough to be a founding member and historian of that alumni chapter the first year it was established.

Eastfield College provided for me, in addition to the academic education, the opportunity to realize that I could accomplish much more. The people there made me feel that I belonged, which in turn led to self-assuredness and an improved self-concept. There I learned to become involved and to give myself the chance to excel. It is that part of my early education that has since enabled me to set and achieve goals and that has encouraged me to continue to want to aspire to achieve more.

After graduation from Eastfield in May 1978 I attended Southwest Texas State University. I majored in social science and received an Award for Excellence in History in the spring of 1980. I graduated with highest honors, obtaining a bachelor's in education in May 1980.

The following August I began teaching at Lewisville High School in Lewisville, TX. I taught there for four years before resigning to pursue a postgraduate education. In May 1988 I graduated from the Texas Tech University School of Law with a doctor of jurisprudence and a master's in public administration.

Audrey L. Lloyd graduated from Eastfield College, a part of the Dallas County Community College District, in 1978 with a degree in social science. She now lives in Dallas, where she serves as assistant city attorney. Prior to assuming this position, she worked with the Fourth Judicial District of Texas as a briefing attorney.

I moved to San Antonio in August 1988 to work with the Court of Appeals for the Fourth Judicial District of Texas as a briefing attorney. That position traditionally being limited to one year, I moved in July 1989 to take a position with the city of Dallas as an assistant city attorney.

Lazaro J. Mur
Miami-Dade Community College, FL

Back in the summer of 1975, after graduating from high school, I was undecided about a career. I decided to go down to the Florida Keys and spend the summer as a lobster fisherman. After a memorable summer, which involved a great deal of work and which really served as an eye-opening experience, I decided to request a counseling session at Miami-Dade Community College, FL, Wolfson Campus.

With a general interest in business and having previously taken an accounting course as part of a high school senior honors program sponsored by Miami-Dade Community College, I decided to pursue accounting as a career. Throughout my two years at Miami-Dade, I had the benefit of not only taking a number of accounting courses, but I was also given the opportunity to work in the accounting laboratory. The experience I received at Miami-Dade helped me complete my accounting education at Florida International University, after which I became a certified public accountant and worked for the multinational accounting firm of Touche Ross & Company as a tax adviser.

Thereafter, I received my master's of laws in taxation, as well as my law degree, from the University of Miami. To this date, I continue to be thankful to Miami-Dade Community College for the guidance I received early in my academic career.

Today, I am associated with the law firm of Fine, Jacobson, Schwartz, Nash, Block & England. I practice in various areas of tax law with special emphasis on international taxation.

I strongly believe that the road to a better future begins in school, and Miami-Dade Community College was there for me at the beginning of that road.

Lazaro J. Mur graduated from the Wolfson Campus of Miami-Dade Community College in 1975 with a degree in pre-accounting and is now living in Miami, FL. He is a member of the Florida Bar and the American, Dade County, and Cuban-American bar associations. He is also a member of the American, Florida and Dade County institutes of certified public accountants and was named in the 1989 edition of Who's Who in American Law.

Linda D. Regenhardt
Northern Virginia Community College, VA

As with most people, the hardest thing for me to do is to change my status. For years I regretted that I had not completed a college education, but instead had gone directly to work, married, and raised five children. I intended to go back and finish my college education, but I somehow never found the incentive to take that first step.

It took a tragedy to get me out of my state of inertia. My best friend started back to college and had encouraged me to do the same. I procrastinated, as my youngest child was only in the first grade. My best friend, along with her eldest child, was killed in an automobile accident in July 1982. The loss of my friend was a reminder of my own mortality. Perhaps I didn't have all the time in the world. It was at that point that I decided to obtain the education I had not pursued. Ten days after my friend's memorial service, I walked into the guidance office at Northern Virginia Community College and signed up for 21 credit hours.

Quite frankly, I was scared to death. I wasn't sure I could think about anything other than maintaining a home and caring for my children. I had been absent from any sort of studies for nearly 16 years.

What I discovered at NVCC was a group of supportive people who really wanted me to succeed and who would do everything in their power to see that I had every opportunity. There was nothing quite like the end of first quarter, when I looked at the posted exam grades and found that I had a perfect 4.0 average. I had proven I could do it.

When it was time for me to transfer to a four-year institution, my adviser from NVCC arranged for me to meet with an adviser from George Mason University to discuss exactly what would or would not transfer. Much to my delight, I found that everything of substance I had taken transferred readily, and I was off to a good start at George Mason. I completed my bachelor's degree in 1984 and moved on to the study of law at Georgetown University, graduating in 1987 with a juris doctorate.

The community college experience was one of such warmth and support that I never severed my ties. During two of my three years of law school, I served as president of the alumni Federation for NVCC and have since that time served on the Educational Foundation Board. I strongly believe that my experience at NVCC gave me what I needed to pursue my educational and professional goals. The community college experience provided a sense of accomplishment and self-assurance, and I was fortunate to be in the company of warm, supportive people who have remained my advisers and friends in the ensuing years. To this day, when I feel the need for advice, I contact former teachers and counselors at NVCC who have proved a never-ending source of support for me.

Linda D. Regenhardt graduated from Northern Virginia Community College in 1982 with a degree in business administration and general studies and is now living in Springfield, VA. She is part of the litigation team for the firm of Clary, Lawrence, Lickstein & Moore. She is licensed to practice law in the Commonwealth of Virginia, the state of Maryland, and the District of Columbia.

Deborah F. Sanders
Columbus State Community College, OH

Before I entered Columbus State Community College, OH, my goal was to become a legal secretary. As a result of the technical training and the individual attention I received at Columbus State, I have had rewarding experiences as a legal secretary and a legal assistant (paralegal). I am now a practicing attorney in my hometown of Columbus, OH.

While attending high school, I didn't dare dream that I could ever join the ranks of the "true" professionals whom I identified as doctors, lawyers, or PhD's. These people were either brilliant or had gone to "the best schools." Although I had always been a good student, I never considered myself brilliant; and because I was educated in inner-city public schools, I was afraid to go to a large college where I would be forced to compete with the brightest students from "the best schools." My dilemma was solved when I received a full scholarship to a community college. I would learn a marketable skill without being forced to adapt to campus life at a large four-year college.

I did not realize my true potential until after I became a student at Columbus State. My academic success gave me confidence that I could compete with students on a larger campus. When I learned of the different careers a legal secretarial position could launch, I began to entertain the idea of becoming an attorney. The study skills I learned at Columbus State carried me through paralegal school, helped me obtain a bachelor's degree, and prepared me well for law school.

The individual attention I received at Columbus State also helped shape my life. During my last year, I considered dropping out of school. One of my instructors counseled me extensively on the benefits of sticking with the program. I don't believe I would have received this kind of personal attention had I been on a large campus. I am sure my life would be very different had it not been for the personal interest of one special instructor.

My education and personal growth at Columbus State brought out the self-confidence to compete with the rest of the world—something I definitely needed to survive the challenge of law school. Because of my experience at Columbus State, I constantly tell young people that they have the potential to do great things. Many students who have reached their senior year of high school without preparing for college feel that it's too late to set high goals. I tell these students that it is never too late to turn what they think is a mediocre existence into a highly meaningful and productive life, as long as they have self-confidence, ambition, and perseverance.

Deborah F. Sanders graduated from Columbus State Community College in 1974 with a degree in its legal secretarial major, AAB. She is now a practicing attorney in her hometown of Columbus, OH, with the firm of Baker & Hostetler.

Nancy E. Slovik
Coastal Carolina Community College, NC

I enrolled at Coastal Carolina Community College, NC, at the age of 17 to earn an adult education diploma. Two years later, I earned my associate of arts degree and graduated with honors. I was well on my way to completing college. I had, however, gained more than an AA degree during those two years.

The school's nurturing atmosphere, my contact with a unique student population, and my involvement in the many college activities available helped me mature and develop my leadership and communication skills. I believe those characteristics were crucial to my acceptance to a "Top Twenty" law school and to my subsequent success there. Moreover, my life is still improving because of my experiences at a community college.

I am especially grateful for the opportunity to have begun my college education at Coastal. After skipping my senior year of high school and beginning college at such an early age, I could have been overwhelmed by a larger institution. I would have felt lost, homesick, and intimidated by too many people. Coastal provided the perfect transition. The instructors knew most students by name. The students tended to really want the education they were getting. I felt comfortable, capable, and confident.

My self-confidence may have fostered a desire to get involved, and I joined the Student Government Association. After a year as secretary, I was elected president of the SGA. In this capacity, I served as an ex-officio member of the board of trustees and as a student representative on several college committees. Other activities, such as membership in the National Spanish Honor Society, kept me busy, but being president of the SGA made me feel responsible—for money, for the success of planned activities, for the reputation of students with the faculty and administration, and for my own reputation. The experience was invaluable.

After completing my studies at Coastal, I transferred to a much larger four-year college, where I graduated magna cum laude with a bachelor of arts degree in English. I then went straight to law school at a university of more than 21,000 students. I obtained a juris doctor degree three years later. Those large schools gave me an excellent education and I enjoyed my years there. However, when I think of Coastal Carolina Community College, I feel especially proud—of the school and of my accomplishments. It was at Coastal that I received the foundation I needed, and it was there that I began to see myself as a successful person.

Nancy E. Slovik graduated from Coastal Carolina Community College in 1983 with an AA degree. She is currently employed as an attorney at the United States Postal Service Law Department in Windsor, CT. She volunteers with the Big Brothers/Big Sisters of Hartford; the confraternity of Christian Doctrine as an instructor for children; St. Joseph's Church, where she serves as a lector; and the Boy Scouts of America as a class instructor. Nancy is also a member of the North Carolina Bar Association.

Roberta L. Stonehill
Ocean County College, NJ

In 1976 I began my transformation from a battered housewife to a professional woman. I had just finished two years of unsuccessful job hunting. Divorce was frequently on my mind, but with three school-aged children, no job, no money, and no help, I felt trapped.

Ocean County College, NJ, was an experiment. My first semester consisted of one course built around my children's schedules and a part-time job. It was years since I had attended school. I wasn't sure I could do college work. In addition, I thought I would feel old and out of place. I found, however, that I could do the work, there were many returning students, and I never felt out of place. I had taken the first steps on the journey that completely changed my life.

I spent four wonderful years at OCC progressing from one course per semester, to two, to three, and onward.

Outside the classroom I was invited to join the honor society and became an active member, then secretary, and finally chapter president. Before long I was involved with many other campus activities; I tutored other students and worked in the library.

Meanwhile, my personal life was changing. I consulted a lawyer about a divorce and concluded that living on welfare would be better than living under the conditions I had endured for so many years in marriage. I was divorced in November 1979. My children and I received welfare for four years.

In 1980 I received my associate in arts degree with honors, was class commencement speaker, received many awards, and was going on to Georgian Court College, NJ, for a four-year degree.

In 1983 I graduated magna cum laude with a 3.87 grade point average, a bachelor of arts in history, and a certification to teach social studies.

Before completing the BA, I decided to earn another degree. I applied to law school. To my surprise, I was accepted by Rutgers University School of Law-Newark, NJ. In 1986, to my amazement, the degree of juris doctor was conferred upon me. I was glad to have witnesses at graduation because I could hardly believe this had happened to me. In 1987 an invitation was extended to me to teach part-time at Georgian Court College.

I realize that OCC provided the opportunity to have a whole new life: to move from a battered housewife without a job or skills to a practicing attorney and college instructor. Whatever contribution I am able to make to OCC and the world can never equal the contribution OCC made to my life.

Roberta L. Stonehill graduated from Ocean County College, NJ, in 1980 with a degree in history. Stonehill is a member of the Ocean County and New Jersey bar associations. She not only teaches history courses at Ocean County College, but also served as a member of the board of directors of the Ocean County Head Start Program. She has volunteered countless hours to the OCC Alumni Association and its efforts to raise funds for scholarships for needy students. She also spends a great deal of time counseling clients at jails at whatever time of day she is needed.

Vivian A. Sye-Payne
Salem Community College, NJ

When I graduated from high school in 1965, career options for Black women were pretty much limited to teaching or nursing, neither of which appealed to me.

By the time I enrolled in Salem Community College, in 1973, I was married and the mother of two children. At that time people like myself were called "nontraditional" students.

Salem Community College had only begun offering a liberal arts curriculum the year before I enrolled. During one semester, I had to attend school five days and two nights per week in order to carry a full courseload. In a two-year period I had one professor for six different courses. That is not a complaint.

The size of the liberal arts faculty was in no way a measurement of their quality. I have often wondered how a small fledgling community college attracted such excellent teachers. They were intellectually stimulating, challenging and encouraging me to be the best that I could be and go as far as I could go. I had one teacher who taught both English composition and American literature. She would never let me be satisfied with a "B" on my papers. She insisted that with a little more work, the paper could rate an "A." I hated her during those times, but there were many days at Rutgers University, NJ, and the University of Pennsylvania that I thanked God for her and her insistence on excellence.

In fact, all of my Salem Community College teachers took a very serious view of their responsibility to prepare my classmates and me to enter four-year institutions. Besides excellent academic preparation, the encouragement I received from my teachers built my self-confidence to the point where I looked forward to new experiences.

After graduating from Salem Community College, I matriculated at Rutgers Camden College of Arts and Sciences, where I earned a bachelor's degree in urban studies and community development. In 1977 I entered law school at the University of Pennsylvania and earned a juris doctorate in 1980. I now maintain my own law practice in Philadelphia. I can honestly say that but for Salem Community College and its wonderful faculty, I would not have pursued such lofty goals.

Vivian A. Sye-Payne graduated from Salem Community College in 1975 with a degree in liberal arts and is now living in Salem, NJ. Her current activities include membership in the American Bar Association, Philadelphia Bar Association's Family Law Section and Probate and Trust Section, and the National Bar Association Women's Division. She is a member of the Salem Community Education Center Advisory Board and founder/chair of the Concerned Black Business and Professional Women's Scholarship, awarded to a college-bound Salem High School graduate.

Hilda Tagle
Del Mar College, TX

From time to time, when I am being introduced and the list of my credentials is being recited, I flash back to the time when I was about 10 years old as I watched a television commercial that said, "Give to the college of your choice, for the college doors will be closing in 10 years." I longed to go to college, but I was so afraid that by the time I was ready, those doors would slam shut in my face.

But Del Mar College, TX, opened its doors for me, and once inside those doors, I found the means by which I could reach heights I never knew existed. Without a single doubt, the knowledge and skills I acquired while I was a student at Del Mar College have served as the educational foundation for what I have achieved in my personal life.

Since my return to Corpus Christi to practice law, my life has been continually influenced by Del Mar. For instance, I have enrolled in a variety of courses, ranging from accounting to data processing, in order to enhance my skills as a lawyer. Each time I have enrolled, I have persuaded one of my four brothers to enroll along with me, thereby gaining mutual benefits: I have acquired an instant study partner, and they have received credit toward a Del Mar degree that they will one day complete.

In addition to honing my skills as a lawyer, Del Mar has provided me with what I consider one of my greatest career achievements, that of being a teacher on the Del Mar College faculty. I learned to appreciate many things, among them the fact that being a good teacher is a great deal harder than I ever dreamed it was going to be. I also learned to appreciate the beauty and determination of older students who, for a myriad of reasons, had either just completed their General Educational Development (high school equivalency) program or had never pursued a degree after high school graduation.

This experience has helped me gain valuable insight, because now that I am a judge, I see juveniles and adults throughout our society who have failed as citizens. I strongly believe that this has happened because of their failure as students. Therefore, I frequently include the completion of a GED as a condition of probation. As a teacher, I have seen improvement in the quality of lives of students who return to school. It is for that reason that I am convinced that this requirement is often the first step of probationers toward becoming productive members of society.

I feel certain that the goal of these probationers can realistically be accomplished using the resources of a community college such as Del Mar College. For this reason, I am thankful as an alumna, a judge, and as an advocate of adult literacy that we have a college of Del Mar's caliber in the Corpus Christi community.

Hilda Tagle graduated from Del Mar College in 1967 and is now living in Corpus Christi, TX. She became the first Hispanic woman in the state of Texas to sit on a court of record when she became the judge of Nueces County Court-at-Law No. 3 in 1985. She was recently appointed by the Texas Supreme Court to the State Commission of Judicial Conduct.

Jeannine A. Thoms
McHenry County College, IL

There is a well-known term in real estate, "Location, location, location." I can unequivocally state that I would not be an attorney today if it were not for the fact that McHenry County College was located to meet my needs. Of course, needs go much deeper than location, and in that regard, MCC was the educational inspiration I needed to consider furthering my goals.

I returned to school as a wife and mother of two young children who depended greatly upon me. My goal at that time was much less than a law degree. The transition to school a mere two miles away was a workable and guiltless one that I could justify. The initial undertaking was also economically feasible for my family. The MCC faculty were available to the students, and their enthusiasm appeared to be authentic. This reinforced my decision to be there.

When I graduated from MCC with high honors, I thought, as many do, that these grades were not indicative of a "real" university. How wrong I was! My accomplishments at MCC carried through to a bachelor's degree, summa cum laude, and a subsequent law degree with high honors. I recite these scholastic achievements not to bore the reader, but to make clear that the standards at MCC were high and that the community college laid the groundwork for me to excel.

After practicing law in a large firm in Chicago's Loop, I decided to return to the community where I live and where my career was launched. I now practice in Crystal Lake and often recommend MCC to my clients who are struggling to determine their future goals. In a general practice, my clients vary from corporate to estate planning. However, I find it particularly gratifying to work through some harrowing experiences with women who are battered or abused and desperately need the legal and social attention of their community.

Jeannine A. Thoms graduated from McHenry County College in 1979 with a degree in pre-law. She is now a practicing lawyer who resides in Crystal Lake, IL . In addition to her pro bono service to indigent and aged individuals through the Legal Services Center at Chicago-Kent College of Law, she serves as the president of the League of Women Voters in the Crystal Lake-Cary area and as a member of both the Steering Committee of the Illinois Court Watching Project and the Women's Advisory Council to the Governor of Illinois. She has served as a delegate to the White House Conference on Families and has received the Junior Chamber of Commerce Certificate of Distinguished Service.

Our society's values are reflected by that which we are willing to tolerate. Fortunately for all, society has initiated laws that no longer tolerate discrimination or abuse based on gender or race. I am proud to practice law in this country and I am proud of an

Rose Vasquez
Black Hawk College-Quad Cities Campus, IL

By the time I got to junior high school, I had fully decided what I wanted to be when I grew up. Television had been a big influence in my life then, and I found that the "Perry Mason" series had sent the most powerful message—study law. Nevertheless, having a dream and realizing a dream are two very different things. As I graduated from high school, it was all too apparent that not everyone shared my confidence in my attaining such a goal.

Clearly, the message from family and friends was "don't rush into anything." Going to a local junior college was a logical step. I could gain a first-rate college education while allowing myself time to think about what it was I really wanted to do. I was convinced that all of the reasons for going to a local junior college were just too positive to pass up. Little did I realize that going to Black Hawk College could mean so much more in terms of educational experience.

College has to prepare students for the real world. It has been my perception that once students have a degree they are expected to hit the ground running if they want to compete successfully in our world. Black Hawk College offered so much more than academics that I knew when I enrolled that I was going to receive a truly well-rounded education. So many of my experiences at the school were important lessons for the real world.

I took all of the pre-law courses that the college had to offer; the professors and the subjects that they taught provided the backbone for my legal education. But just as important were my other experiences at the college. During my sophomore year, I was president of the student government association. During that same year the association provided the school with its first rock concert. I was also appointed by the president of the college to be on its affirmative action task force. For two years I was on the school's debating team, and in my last year I traveled with three other team members to compete in the annual Harvard University Invitational Forensic Tournament. Looking back, I see that these and other experiences I had at the college provided me with invaluable lessons that I use every day, and not just in the practice of law.

I am currently practicing law as an assistant attorney general for the Iowa Department of Justice. I am an alternate for the Democratic Party's state central committee, and I am an adjunct professor at Drake University Law School. In my opinion, this work is clearly a reflection of my interests and activities at Black Hawk College.

> *Rose Vasquez graduated from Black Hawk College-Quad Cities Campus in 1974 with a pre-law degree and now lives in Des Moines, IA. She is an assistant attorney general for the state of Iowa. She is an alternate for the Democratic Party's state central committee and an adjunct professor at Drake University Law School. She represents the Iowa Department of Health at all levels of litigation.*

David V. Williams
Patrick Henry Community College, VA

Going into my senior year of high school, I knew that I wanted to go on to college. I did not know what I wanted to study, nor did I know where I wanted to go to college.

As graduation approached, my indecision grew. My high school English teacher had on her own secured for me a scholarship at a small North Carolina college, but even with a scholarship I knew college was going to be a financial hardship on my family.

With that in mind, a friend suggested that I go to Patrick Henry Community College and talk to one of the guidance counselors. Ken Lemons, the counselor, pointed out the low cost of PHCC and the fact that the courses I took would transfer to a four-year institution. I decided to go to PHCC, and I have never regretted it.

When spring came and classes started, I was in for a rude awakening. I had done well in high school, and I was not accustomed to making "C"s and "D"s. I realized that in some respects I was totally unprepared to do college-level work. The shock was so great that had I been at a large school where I was just a number, I may very well have quit. At Patrick Henry Community College, the professors were more than willing to help you if you were willing to work. So with a little hard work on my part and a great deal of help on their part, my grades began to rise, and I eventually graduated with honors with an associate degree in education.

Armed with the knowledge that I could do college-level work if I worked at it, I went on to obtain my bachelor's degree in psychology from Virginia Tech and then my JD degree from Campbell University.

I will be forever grateful to the professors at PHCC for teaching me two valuable lessons. One, there is no substitute for work, and two, if you are willing to work, there will always be people willing to help you.

David V. Williams graduated from Patrick Henry Community College in 1975 with a degree in education and now lives in Martinsville, V.A. He is the Commonwealth's attorney for Henry County and was a charter member of the college foundation's board of directors. He has just been nominated by the local bar association for appointment as a circuit court judge.

CHAPTER FOUR

Science, Technology, and Agriculture

Ann Bailey

University of Minnesota Technical College-Crookston, MN

A horseback ride taken during my senior year in high school put me on the road to the University of Minnesota-Crookston and eventually a career in agricultural journalism.

During that summer between my junior and senior years of high school, I thought I had my post-high school plans mapped out. I would attend a four-year university with a journalism program and find a job at a newspaper. That was before a local university instructor brought her daughter to my parents' farm for a little horseback riding. In the midst of our ride, Myrna commented on my fondness for horses. She suggested I look into the light horse management program at UMC.

I hadn't thought about attending UMC because I didn't know it had such a program, and besides, I thought I knew exactly what I would be doing after high school. I was headed to a journalism program at a four-year university. But despite this conviction, I found myself thinking more and more about the light horse management program at UMC. After talking to my parents, I decided to put my journalism plans on hold and to attend UMC.

During my two years at UMC, I not only learned about horses, but I also laid a strong foundation for what was to be a career in agricultural journalism and got excellent preparation for my university studies. The smaller class sizes at UMC allowed the teachers to offer help in my science and math classes, two of my most difficult subjects in high school.

UMC gave me the chance to develop my study skills and taught me responsibility and how to be a self-starter—important tools for my university studies. One highlight of my sojourn at UMC was the opportunity to train and care for a horse for nine weeks. Obviously, the experience built confidence in my abilities to handle such important responsibilities.

After graduation from UMC, I spent a year working at my parents' farm with ample opportunity to use the horse management skills. I then attended a four-year university, earning a bachelor's degree in English. During my last year at the university, I landed a job at the *Grand Forks Herald*, rising from a clerk to a copy editor and a reporter. When a job opened at *Agweek*, an agricultural marketing magazine published by the *Herald*, I found work that would use my agricultural classes and English training.

I cover commodity markets, food safety issues, and agricultural education for the magazine, which circulates in Montana, Minnesota, and the Dakotas. I even manage to sneak in a few stories about that first love—horses—that drew me to UMC. I'm grateful for that horseback ride that helped me onto the right career track.

Ann Bailey graduated from the University of Minnesota in 1979 with a degree in light horse management and now lives in Grand Forks, ND. Through her work in agricultural journalism, she helps make farmers aware of many things, from how to sell corn to understanding women's changing roles on the farm.

David L. Bayler
Olney Central College, IL

Attending Olney Central College was a natural thing for me to do for several reasons.

When my father retired, he moved his family to a community college town because of the excellent opportunities it provided. He hoped we would take advantage of it, so we did.

After taking the math-chemistry-physics route through high school, engineering was the way to go for me. I had aspired to be an engineer for years, since Dad had worked for the Army Corps of Engineers for 37 years.

Olney Central College has a pre-engineering program that parallels the University of Illinois, which had the number-three ranked engineering school in the country. What an opportunity! I decided to attend summer classes in order to graduate with the associate of science degree in 1973. I felt it was important to obtain the AS degree in case my goal of a BS degree fell through.

Olney Central is where several of my friends were going. It was an enjoyable two years. I also met several new people and made lasting friendships. Several of us had the same classes together for two years.

Another advantage at OCC was that we had the same teachers for several classes. The teachers knew what we could do, and we knew what to expect from them. OCC had excellent, dedicated teachers who cared enough to give us extra help when we needed it. I needed it a few times! I am thoroughly convinced that without the extra attention of a few special teachers at OCC my career as an engineer would have ended before it ever started. Even through graduation from Eastern Illinois University with an MBA in 1985, no teacher had a more positive motivational effect on me than Ray Culver at OCC.

After transferring to the University of Illinois, I discovered the classes that I took at OCC were the same courses that the U of I used to weed out the freshmen and sophomores. I was glad that I had those classes out of the way.

The economic benefit of a community college was another plus. I worked part-time and could pay for the low tuition and book fees myself. I know my dad appreciated not having to finance those same two years at the university.

I am glad to have had the opportunity to attend OCC and tell anyone who will listen that it was definitely a good way to start a new career. Speaking to high school students, I encourage them to investigate their local community colleges. It could possibly be the natural thing for them to do, too.

David L. Bayler graduated from Olney Central College in 1973 with a degree in engineering and now lives in Paris, IL. He works for the Bureau of Construction of District 5 of the Illinois Department of Transportation as a civil engineer and supervising field engineer. He is a trustee for the local chapter of the Illinois Association of Highway Engineers and vice chairman of the board of The Paris Highway Credit Union.

Shahzad Bhatti
Kansas College of Technology, KS

The sight of a tree in full bloom is magnificent. The branches stretch out, like huge arms, reaching out for the sky. The leaves, drenched in sunlight, dance in the breeze. Juicy colorful fruit hangs from its branches like Christmas ornaments. However, all its growth and prosperity depends on its roots.

My growth and prosperity also depend on my roots. Part of my roots are entrenched at Kansas College of Technology. In 1982, I received an associate of technology degree in computer science. The education I received was such a bargain that I could not resist continuing for another year and earning two more associate degrees, one in computer engineering and the other in electronic data processing.

Two weeks before I graduated, NCR Corporation interviewed on campus and offered me a job. I went on to receive a bachelor of science in computer science, and a master of science in computer engineering with NCR's financial support. My education and training at Kansas College of Technology laid a foundation upon which I was able to build not only a technical career, but also an educational career.

In retrospect, Kansas College of Technology was the best thing that happened to me for the following reasons:

- The quality of education was excellent
- The cost of education was a fraction of the cost at other colleges
- In two years I was able to earn my first degree
- The college helped me find my first job

Despite all these tangible benefits, what I treasure most from the Kansas Tech experience is the friendship of Robert Homolka (calculus professor), the love of Les Kinsler (physics professor), the inspiration of David Delker (digital design professor) and the affection of Anthony L. Tilmans (college president). Thanks, Kansas Tech!

Shahzad (Shah) Bhatti graduated from Kansas College of Technology in 1983 with degrees in computer science, computer engineering, and electronic data processing and now lives in Longmont, CO. He is a project engineer at Solbourne Computer, Inc. He is also a lecturer for Wichita State University and was a panelist in a workshop by the American Association of Artificial Intelligence.

David L. Boprie
Washtenaw Community College, MI

To graduate from college was one of many goals for myself and would also be a first for anyone in my family. After high school, I was not ready for college and enlisted in the Air Force for four years. That was a great time, especially the travel. I never lost sight of my original goal, however, and did some basic coursework at colleges or universities where I was stationed.

Winter of 1978 was my first semester at Washtenaw Community College. I found the flexible hours at WCC fit well with my full-time work schedule. One refreshing discovery was mathematics. The math lab made learning fun, and now I had a great application, electronics.

While gaining work experience, I completed associate degrees in general studies, electronics engineering technology, and digital technology. Washtenaw has given me the basic tools that have allowed me to do well in a wide variety of electronics-related work. My background now includes programmable controllers, nuclear imaging, machine vision, and aerospace systems.

Currently, I work at the University of Michigan Space Physics Research Laboratory and enjoy it very much. We have recently completed the integration of a High Resolution Doppler Imager (HRDI) aboard the Upper Atmospheric Research Satellite (UARS). HRDI is one of nine instruments aboard UARS, all of which are designed to collect data and help us understand the upper atmosphere and its effect on global climate, especially ozone. It will be shuttle-launched in late 1991. I have been involved in many facets of this exciting project, including electronics fabrication, design, testing, and field support.

WCC in combination with my work experience has prepared me with a well-rounded base to try almost any challenge in my field. Many thanks to the great instructors I was fortunate to have. I look forward to returning to the math lab and continuing toward a four-year engineering degree, or maybe just for the fun of it!

David L. Boprie graduated from Washtenaw Community College in 1986 with degrees in digital technology, electronics engineering technology, and general studies and now lives in Ypsilanti, MI. His current occupation is space physics research technician, where he builds instrumentation that helps the scientific community better understand our planet's atmosphere and provides information for policy decisions of the future.

John L. Duncan
Wake Technical Community College, NC

After studying mechanical engineering at a local four-year university for three semesters, I decided to join the Marine Corps. Specializing in aviation showed me I was a hands-on person whose future success lay in a technical education. Knowing that I was going to have to live on G.I. Bill funds when I returned to school, I selected Wake Technical Community College (then called W.W. Holding Technical Institute), both for its affordability and its technical courses. Because of my years of experience in aviation electronics in the Marine Corps, Wake Tech let me start in the second quarter. I remember receiving my first test score—in the low forties—and realizing that this school was going to be no pushover.

I married the following July, and together my wife and I got me through the next two years. I took advantage of the college's cooperative program with local businesses and worked every other quarter. This not only helped financially, but it was also a good introduction to the world of business and provided a break from school.

The engineering associate degree I received in 1969 led to employment at some of the major corporations in the area: Xerox, IBM, and Simplex Time Recorder. In 1986 I joined Triangle Research and Development Corporation as a technician on a variety of government contract and research projects. Over the last several years, I have watched this small business grow and have seen several of my projects evolve from ideas into products. I have assisted on cutting-edge projects for NASA, the Department of Defense, and the Public Health Service that have ranged from cancer research and rehabilitation devices to the development of systems for space flight. I have moved from solely providing technical support to becoming principal investigator on my own projects in the fields of aviation electronics and fiber optics.

My first electoral experience was with the student council at Wake Tech, where I won the secretary/treasury seat by a narrow margin. In 1981 I ran for the Cary, NC Town Council and was defeated; however, in 1989 my bid was successful. With the help of Wake Tech graduate Paul Smith and his wife, Addie, I defeated a 14-year incumbent by six votes.

A community college such as Wake Tech offers not only affordable education, but also quality courses to individuals wanting to gain practical knowledge and enter the work force after two years of study. The co-op plan with industry allows students to gain experience in the workplace. In some cases, students may need to continue their education, while in others they are fully qualified to begin work. In my case, I will always hunger for more, thanks to Wake Technical Community College.

John L. Duncan graduated from Wake Technical Community College in 1969 with a degree in electronics engineering technology and now lives in Cary, NC. John is project engineer for Triangle Research and Development. As a technical researcher, he has furthered the development of many products and systems to help the disabled, elderly, and handicapped. He has been elected to the town council of Cary, NC, where it is his goal to help maintain the quality of life in one of the fastest-growing small cities in the South.

K.G. Engelhardt
Ohlone College, CA

Ohlone College has provided me the historically unprecedented opportunity of becoming a lifelong learner. It gave me the basic preparation and motivation to go on to academic and professional success at Stanford and Carnegie Mellon universities. This was not trivial. I entered Ohlone after I had three teenage children and after being out of school for many years. But, more importantly, Ohlone gave me a renewed confidence in my own ability to continue to learn and grow throughout my life. I am presently residing in Pennsylvania and continue to take classes at community colleges. I have the privilege of selecting any course I need for self-enhancement or for increasing my knowledge base in unexplored domains. It is this quest for knowledge that has led me into my career in robotics. Robotics is one field of technology that will continue to expand into the 21st century.

In our present culture, it is estimated that 50 percent of our technical knowledge becomes obsolete about every four years. If we are to remain a leading technological nation, we must encourage a focus on lifelong involvement in the learning process. This is the invaluable opportunity our community colleges provide us.

It is this opportunity to begin again, better yet, to continue to review learning throughout our lifetimes, that makes the community college system unique and vitally important to our globally competitive America. Education is a privilege. Community colleges are, indeed, unique entities, providers of this unique privilege.

Age is no longer a barrier to educational opportunities. This nation NEEDS mature as well as young persons thinking, learning, and making decisions together. That is what the community colleges' lifelong learning provides. Community college education must be protected and enhanced. Its value cannot be overemphasized. I take great pride in my community college roots. I remain enormously enthusiastic about this most unusual educational experiment.

K.G. Engelhardt graduated from Ohlone College in 1977 with a degree in liberal arts and now lives in Pittsburgh, PA. Her current position is director of the Center for Human Service Robotics for the Mellon Institute. She is also the founder and current director of the Center for Human Service Robotics at Carnegie Mellon University. Engelhardt has served as advisor to projects of the National Institutes of Health and NASA, and in 1989 she received the Oustanding Woman of Pittsburgh Award.

Janet Anderson Glynn
Technical College of the Lowcountry, SC

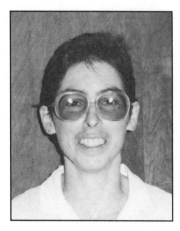

Before college, my main purpose in life was to be a good mother for my son. Then I began to wonder what I would do when he was grown and on his own. I decided to seek the answer early on, hoping that I would have something for myself when that event occurred.

Upon enrolling at the Technical College of the Lowcountry, I began a two-year journey that became almost as rewarding as being a mother. I rediscovered many hidden characteristics within myself that I hadn't used since high school (15 years earlier). Self-discipline was one lesson I had to learn quickly. As a mother, I hardly used it except to restrain myself from buying every toy in the store for my son. In school I was expected to meet deadlines and see projects through to the finish or suffer the consequences. The instructors at TCL helped me accept my new responsibilities and offered suggestions to lessen the load.

Self-confidence was a bit harder to achieve. I was fortunate to have had enthusiastic instructors who challenged me to search for the answers to my own questions and gave me the opportunity to improve and display my abilities. Their faith in me made it much easier for me to begin believing that I could do just about anything.

Upon graduation I reached the ultimate goal—self-respect. I am proud to say that I couldn't have done it without my new family at the Technical College of the Lowcountry. The knowledge I gained in my two-year journey could not be taught from a book. The caring and understanding that these people expressed helped me to reach many goals and regain parts of me that had been lost for years. And it also let me find out early in motherhood that my 10-year-old son was very proud of me, too.

Janet Anderson Glynn graduated from Technical College of the Lowcountry with a degree in computer data processing and now lives in Beaufort, SC. When she graduated in 1989 she was a student speaker and graduated with honors and as a member of Phi Theta Kappa. Now she works in her chosen field as a data entry clerk for Mim, Trask & Co. P.A.

David Harlan

The Ohio State University Agricultural Technical Institute, OH

Five years after graduating from high school and being married (with a new son), I still did not know what I wanted my profession to be.

After preparing for college in high school, I instead opted to go into my father's business and start making money immediately. After three years, I decided this was not the way I wanted to spend the rest of my life.

I held a series of positions after leaving my dad's business—from truck driver to factory worker—without much satisfaction in any of them. Then, one day, I had the good fortune to run into my high school guidance counselor. After some discussion of my employment history, likes and dislikes, and special interests, he suggested The Ohio State University Agricultural Technical Institute in Wooster, OH. OSU/ATI was a fairly new school, designed to provide an education and work experience through a practicum/internship program in one's course of study.

I have always liked the outdoors and thought there is a need to protect our natural resources. OSU/ATI's Soil and Water Conservation Program seemed to be the perfect field of study for me.

I visited OSU/ATI in 1976 and met with the curriculum adviser and discussed courses, fees, job opportunities, etc. I decided this was what I wanted to pursue as a career. Here was a chance to do something I believed in and help protect our environment at the same time. After seeing the OSU/ATI campus, I realized that those enrolled were not all fresh out of high school. There were people my age (25), too. Some had degrees from other universities but were here at OSU/ATI taking some of the specialized courses only this school had to offer. Others were like me—finally deciding what they wanted to do in life.

I fit in immediately at OSU/ATI. The instructors were helpful, supportive, and extremely interested in someone who had forgotten many of the things he had learned in high school five years earlier. It was easy, too, at a smaller campus to become familiar with the faces and names of the staff and students. There was an atmosphere that made me want to learn, and the desire to learn took hold. The courses and internship helped prepare me for a career in soil and water conservation, which I started immediately after graduation and am currently in my twelfth year. I am not getting rich, but the rewards of the job are a fortune in themselves, and I enjoy my work immensely.

Thanks to OSU/ATI, I have a profession I am proud of and which I believe, over time, will contribute to the protection and enhancement of our soil and water resources.

David Harlan graduated from The Ohio State University Agricultural Technical Institute in 1978 with a degree in soil conservation and lives in Oak Harbor, OH, where he holds the title of conservationist.

Allen Harper
Allen County Community College, KS

Perhaps my participation in Allen County Community College activities and my graduation with an associate in science degree involves one unique circumstance. I believe that, at the age of 81, I am the oldest graduate of the college.

After having obtained a bachelor of science in pharmacy from Kansas University in 1931, I decided to resume studies to update my knowledge of soil conservation and market gardening. Informative courses with Don Benjamin and Leslie Meredith brought me up to date on these levels. Courses in small business management and marketing under David Roos rounded out the entire production-marketing program.

I moved from Moran this past July to take residence with my son Jim and his family. They purchased a home and small acreage five miles west of Iola, having sold their town residence to the Iola School Board for conversion to a high school parking lot.

In addition to Jim and his wife, Nova, and me, their son, Chuck, and daughter, Patience, compose the household. An assortment of small creatures includes four cats, four hens, and a dog.

The acreage is divided into two sections, one devoted to a commercial planting of irises and daylilies, and a second to a sizeable vegetable garden from which items are frequently taken to the Iola Farmers' Market. An interesting herb garden is one of Nova's favorite plantings.

My interest in irises dates back some 40 years, beginning in Kansas City and continuing there until 1974, when I moved to Moran. It gradually developed into a commercial venture, continuing to the present.

Our iris and daylily venture is mostly a mail-order operation. We issue a catalog mailed chiefly to iris lovers in the four-state area of Kansas, Missouri, Arkansas, and Oklahoma. Orders are generally processed and shipped during July and August, which are the ideal months for planting. We do on occasion receive orders outside the designated area, as evidenced by our first foreign order, one from Krefeld, West Germany.

We are still moving plants from our Moran garden, and so far have reset over 3,000 plants. Jim and I both hybridize and have several varieties registered with the American Iris Society, of which we are both life members.

I definitely believe that my studies at ACCC have contributed positively to the distinct improvement of our business methods and the conservation of our natural resources, and I am most grateful to Allen County Community College for its contribution toward bringing this about.

Allen Harper graduated from Allen County Community College in 1989 with an associate of science degree. Allen, a former federal employee, is the oldest graduate of ACCC, and was active with intramurals, community theater, and was an American history tutor while at the college. He is currently an active horticulturalist.

Melina Huddy
Hocking Technical College, OH

Were I to meet you for the first time today, I would introduce myself and say that I live in Newark, DE, where I work for Lanxide Corporation as a technician. Depending upon your interest, I might go on to explain that Lanxide is a research and development firm in the advanced ceramics industry and that I work in the Industrial Products Division. My position requires an associate degree, which I obtained from Hocking Technical College at Nelsonville, OH.

Had we met a short three years ago, late in the summer of 1986, mine would have been a much different story to tell, one not offered so freely at a first introduction.

At that time I lived in Little Hocking, OH, on a farm where I fed livestock in exchange for free rent. My only income was from welfare in the form of Aid for Families with Dependent Children (AFDC) and food stamps. The man in my life was an abusive alcoholic. I had dropped out of school in 1972.

It is not by coincidence that my life has changed so dramatically in such a short period of time. Nothing short of a series of miracles, combined with hard work and a great deal of support from a variety of people, made the transformation possible.

At Hocking Technical College, I found state and federal funds were available for my education. I found I qualified for a Pell Grant as well as Title III money. My tuition was paid, as were all my book fees and supplies. I continued to receive AFDC and food stamps while I studied.

First, I obtained my high school equivalency certificate. Then I was given a scholarship to continue my schooling at Hocking Technical College. I enrolled in the Ceramic Engineering Technology Program and was a dean's list student each of the six quarters I was there.

I was inducted into Phi Theta Kappa and was later named to the President's List, the top 2 percent of the student body. I was also the recipient of the prestigious Trustees' Award, given annually to an outstanding student.

Graduation day was the happiest day of my life, as family and friends traveled some distance to share my special day. A close second for my happiest day was the day I called the welfare office to tell them I was no longer in need of their assistance.

To say Hocking Technical College changed my life is at best inadequate. More precisely, it saved me. It saved me from poverty, from abuse, and from a lifetime of fear. My heroes are many, including the entire engineering department and each and every one of my instructors. My best friends are my classmates, and we stay in touch.

When I count my blessings, right after my kind and gentle husband and my happy and healthy children, I thank God for HTC and the knowledge I received there. Beyond the science and math and the classroom learning, I discovered myself and the life I now have.

Melina Huddy graduated from Hocking Technical College in 1988 with a degree in ceramic engineering technology and now lives in Newark, DE. She is a technician for Lanxide Corporation.

Barry Jacobs
Central Community College, NE

The education I received at Central Community College was instrumental to my success today. I received an associate of applied science degree and found this a worthwhile achievement.

One of the things I enjoy about the welding profession is that there are two aspects to it: one to physically perform the work, and the other to study and engineer and direct the work.

After graduating from Central, I became a welding teacher at Hobart Technical School in Troy, OH, and found that my knowledge and welding skills were as good as those of other teachers with more hands-on experience. I found that two years of formal education had taught me what it would have taken many years of job experience to learn.

After teaching, I had an opportunity to work in the nuclear power industry—the high-tech area in welding—and decided I couldn't turn down this opportunity. I started at the bottom as a welding technician, using semi-automatic welding equipment, doing procedure studies, and learning new welding codes. After becoming proficient in semi-automatic welding, I became a welding engineer and enjoyed the next several years as I worked with codes, procedures, and new techniques of welding in the nuclear power industry.

I later was hired as a welding engineer by Morrison-Knudsen of Boise, ID, because of my nuclear welding experience. I was later promoted to senior engineer, overseeing welding procedures, nuclear-related requirements, and code requirements. The company later transferred me to a nuclear plant that was being overhauled and promoted me to project welding engineer. I supervised eight welding engineers and 63 welders on the $90 million project, which we brought in ahead of schedule and under budget. In the process, we developed new welding techniques that are now standard in the nuclear power industry, saving consumers millions of dollars on nuclear rework projects.

My profession has given me the opportunity to travel throughout the country, working on and serving as a welding consultant to nuclear power plants. After leaving the nuclear industry, I opened my own welding fabrication business, and I'm currently planning for a 300 percent expansion in the spring of 1990. I can't express my appreciation enough for a fine technical school such as Central and the fine instructors who made this all possible for me.

Barry Jacobs graduated from Central Community College in 1977 with a degree in welding technology and now lives in Mt. Vernon, OH. He is currently owner and operator of Welding Fabrication Shop. He is a member of the Elks and sits on a leadership committee for the local Boy Scouts. His company sponsors a bowling team, a girls' softball team, and a boys' softball team.

Marshall G. Jones
Mohawk Valley Community College, NY

While I was in high school, I thought that my only ticket to college would be via a sports scholarship, because I was only an average student. I was always an above-average mathematics and science student, and below-average in the remaining core courses. But average grades were not enough for acceptance into an engineering school in New York state during the late '50s and early '60s. The New York engineering schools were all private at that time and were very expensive. Again, the route to college for me appeared to require having a sports scholarship. This strategy was on track until I sustained a sports-related injury during my junior year. My dreams of college were shattered. It was then that my guidance counselor advised me that perhaps a community college would be a good choice.

I took my counselor's advice and started looking for a school that offered the mathematics, science, and mechanical drawing that I enjoyed. After considering several schools, I chose Mohawk Valley Technical Institute (now Mohawk Valley Community College) because it appeared to have the best program (mechanical engineering technology) for me.

I was very apprehensive about going to college because of the fear of not succeeding. But MVCC provided the atmosphere I needed. The atmosphere, and my desire not to fail, resulted in a successful career at MVCC—a success that opened the doors to many other opportunities.

It was only because of my performance at MVCC that I was able to transfer to the University of Michigan's mechanical engineering program, rated fourth in the nation at the time. It was at MVCC that I first realized that I can accomplish almost anything if I am willing to work hard. It was after I finished Michigan with a BS in mechanical engineering, and had worked for four years, that I set my goals a little higher and decided to pursue a graduate education at the University of Massachusetts. It was there that I received my MS and doctorate in mechanical engineering. Note that it was the early laboratory skills learned at MVCC that helped me to be successful in graduate school as well as at General Electric Corporate Research & Development today.

My MVCC experience showed me that dreams can come true if I am willing to work for them. I am truly proud that I attended MVCC, and I am a true advocate of the community college system. For this reason, I tell students at all levels, and their parents, that the community college route to an education is a viable choice.

Thank you, MVCC, for giving me that viable choice.

Marshall G. Jones graduated from Mohawk Valley Community College in 1962 with a degree in mechanical engineering technology and now lives in Scotia, NY. He is senior research engineer and project leader in laser technology for corporate research and development at General Electric Company. He was the recipient of an honorary doctor of science degree from the State University of New York and received the 1986 Distinguished Engineering Alumnus Award from the University of Massachusetts.

Elizabeth M. Kendall
Portland Community College, OR

Recently, my niece commented, "Aunt Betty, I'm so proud of you. Look how far you've come." The comparison she drew between 15 years ago and now came as a real surprise to me. I believed the feelings of futility and self-doubt under which I had labored so many years ago were well concealed behind the facade of a middle-aged woman who was a successful housewife, mother, and community leader. No one, I was convinced, suspected the amount of courage it took for such a person to enroll in a course for professional auto mechanics—especially in those days.

My journey began with the acquisition of a new car—one with a talent for breaking down in places where repair facilities were unavailable. My amateur repairs were successful, but I remained convinced that any other individual would have coped much more ably. Determined to stop being a royal pain to myself and everyone else, I enrolled in the automotive repair technology course at Portland Community College, expecting to survive only long enough to achieve a modicum of automotive literacy.

The instructors had other ideas. Although unaccustomed to having women in their classes, they were committed to the progress and professional achievement of each individual student. They were demanding, but they were always available for counsel and assistance. I realized my curiosity about auto mechanics had been deeply repressed for a lifetime, and I developed a strong determination that others should not be in that situation. This feeling is with me today and is also one which exemplifies the mission of Portland Community College.

My "modicum of automotive literacy" stretched into a two-year associate degree in automotive repair technology. I was graduated, went to work as a mechanic, and became certified by the National Institute for Automotive Service Excellence as a Master Technician.

Three years later, I accepted a teaching position at PCC. I was encouraged into yet another profession, one which is the most stimulating and demanding of all. My professional world expanded. I became vice president of the Oregon Vocational Trade Technical Association and was elected to the governing board of the Oregon Section, Society of Automotive Engineers. In 1987 two other repair technicians and I formed the nucleus of the Association of Women In Automotive Related Businesses, which now has a large membership.

I have been teaching for 10 years. The days are frantically busy, and sometimes the old feelings return—but only fleetingly! With another vehicle to repair, student to reassure, or lessons to plan, there just isn't room for self-deprecation. Besides, the instructors who knew better than I what I was able to accomplish are now my colleagues. As with my niece, I won't let them down!

Elizabeth M. (Betty) Kendall graduated from Portland Community College in 1976 with a degree in automotive repair technology and now lives in Portland, OR. She is an instructor in the Automotive Repair Technology Department at Portland Community College. She began her automotive experiences as one of a handful of women enrolled as students in the college's two-year program.

Charles D. Kesner
Northwestern Michigan College, MI

A four-year college degree was not in my plans during high school. Instead, I went to Michigan State University to complete an intensive one-year horticultural study schedule called a "Short Course," later known as the "Ag Tech" program. After this, I entered into a partnership with my father on a fruit and dairy farm operation. During the next two years, I developed a keen interest in the fruit-growing aspect of farming and an equal lack of interest in the dairy operation. My plans to increase the former and phase out the latter were interrupted by the Korean War.

During my two years of service, I began contemplating a professional career in horticulture. Some fairly serious problems faced me: I had not taken enough math in high school and practically no chemistry. Returning home, I talked with counselors at the local junior college, Northwestern Michigan College, and found my fears were valid. I would have to take high school algebra and geometry before becoming qualified to enroll at NMC.

Fortunately, these high school classes were available during the evenings, and I could take my college courses at the same time. In fact, my college algebra teacher turned out to be my high school algebra teacher—and he was outstanding. With a great deal of study time both nights and weekends, I made it through high school algebra, high school geometry, college algebra, and chemistry without serious problems.

It's interesting to reflect on my early fears about not having a high school math or chemistry background. By the time I had finished bachelor's, master's, and doctoral programs, I had successfully completed college math, inorganic, organic, and general biochemistry, and plant biochemistry. I have to attribute much of this to the understanding and encouragement given to me back at NMC. I sincerely believe that many students with great potential never enter college because they believe themselves unqualified. This is where community colleges can provide invaluable service, as I believe NMC has done over the years.

Since receiving a doctorate, I have worked at the university level in research, education, and finally as the director of a horticultural research facility. I wonder where I would be if it were not for NMC and its dedicated staff.

> *Charles D. Kesner graduated from Northwestern Michigan College in 1959 with a degree in horticulture and now lives in Suttons Bay, MI. As director of the Northwest Michigan Horticulture Research Station, he is responsible for one of 137 field research stations operated by Michigan State University and has done significant work in orchard systems, trickle irrigation, growth regulation, orchard sprayers, and strawberry harvesters. He has published more than 100 papers detailing his research.*

Harriet Kronick
Bucks County Community College, PA

Returning to school as a mature adult definitely ranks as one of the most crucial goals I have ever made. Frankly, the decision was made mainly for the fulfillment of a void in my life, the acquisition of a college diploma. Having devoted most of my work life to the healthcare profession, I naturally opted to advance my formal education in this area. Honestly, I chose Bucks County Community College because of its location and financial scale. I ultimately received my degree in psychology, life-skills emphasis. I never returned to the profession.

What I encountered at Bucks County Community College was a learning institution geared for today's motivated student, regardless of age. It stands out as an academic center that not only offers a stimulating traditional format, but also is committed to offering a variety of nontraditional academic experiences. If not for this declaration to foster progressive experimental classes, I would not have realized my personal renaissance.

I was encouraged by my academic adviser to step outside the psychology arena when choosing elective classes in an attempt to begin to nurture personal interests. Having always admired the lifestyle of archaeologists, because of what they do and how they do it, I signed up for an introductory historical archaeology course. I looked forward to the class with a mixture of curiosity and excitement. I have never lost my enthusiasm in either of the areas. I was hooked. It did indeed become my lifestyle. My determination to carve a place for myself in this profession was always encouraged by my professors. They remain a source of inspiration to this day.

I remain associated with the college by volunteering my assistance to current archaeology projects that reach out to the community in an effort to nurture interest in our cultural heritage and physical environment.

Among the most outstanding of these projects is the organization of the Bucks County Community College Archaeology Club. Through this college-based arena, I, as founding president, have fostered community awareness and concern among all age groups for the preservation of our past. I am hopeful that my accomplishments will encourage others to pursue new interests. I have shared my successes with the students of Bucks County Community College and nearby schools through lectures and workshop programs.

At present, I am employed as a material culture analyst with a local archaeology firm. But the continued interest and encouragement of one professor for a former student resulted in my acceptance into a college that will afford me the opportunity to receive a bachelor of arts degree in history/social science with a concentration in archaeology.

Harriet Kronick graduated in 1988 from Bucks County Community College with a degree in archaeology and now lives in Langhorne, PA. Her current occupation is material culture analyst. She is the founder of the Bucks County Archaeology Club.

To the board of directors and faculty of Bucks County Community College, thank you for caring.

Ray A. Lattimore
Greenville Technical College, SC

My years at Greenville Technical College were some of the most inspiring years of my life. When I graduated in 1984 with a double major in computer science and marketing, I had been amply prepared and groomed to take a stance in the real world. With my career as a systems analyst at Metropolitan Life Insurance Company off to a booming start, I began to focus some of my attention on community and civic involvement.

Presently, I serve as chairman of the Greenville County Community Housing Resource Board and vice chairman of the Greenville County Democratic Party. In addition, I am a board member and past vice president of the Greenville Technical College Alumni Association. Other appointments include Big Buddies Association of the Children's Hospital, City of Greenville Affirmative Action Committee, and the Keep Greenville Clean Committee.

Upon establishing membership in the Mu Pi Chapter of Omega Psi Phi Fraternity, Inc., I became the assistant keeper of records and seals, chairman of the scholarship committee, and chairman of the social action committee. Still yearning for knowledge and increased awareness, I obtained a real estate license and became a member of the Greenville Board of Realtors, South Carolina Board of Realtors, and the National Board of Realtors. Based upon these accomplishments, I was listed in *Who's Who Among Young Men of America* in 1988.

I truly feel that Greenville Technical College played a major part in all that I have been able to achieve and return to the Greenville community. I am proud to say that I am not the only one who has grown and blossomed in recent years. Greenville Technical College has experienced tremendous growth, and it stands as a pillar and image-maker in the community. I am doubly proud to say that I have returned to my alma mater as a part-time instructor in the Computer Technology Department. I will be forever grateful to Greenville Technical College, and I highly recommend this institution to anyone who has the desire to grow and achieve personally and professionally. Hats off to Greenville Technical College!

> *Ray A. Lattimore graduated from Greenville Technical College in 1984 with a degree in computer science and marketing and now lives in Greenville, SC. He is both a goodwill ambassador for community-based education and a much-sought-after leader in the Greenville community. His current occupation is systems analyst.*

Caroline Laudig
Indiana Vocational Technical College, IN

We all have a measure of success in our lives, and sometimes we must be reminded that we are successful. My reminder was the result of my re-education at a local technical college. I had a college preparatory high school education and several years of college, which prepared me for a career in a highly technical industry. A major change in my personal life allowed me to consider a new career in computer programming.

I realized that I would need immediate specialized training and direct experience to take advantage of this opportunity, and Indiana Vocational Technical College allowed me to focus on this area without the distractions of filling electives in different areas. While gaining new skills was a major goal of this course of study, feeling successful at learning new things was a secondary result from concentrating on a particular skill.

The success that I had experienced in school has given me the confidence to undertake a project that still occupies me spiritually, physically, and emotionally after three years—the total renovation and rehabilitation of an historic house. This house, significant for its architectural style and location, has become a landmark project for inspiring others to save houses that would otherwise have been destroyed.

Another aspect of success in professional and personal endeavors is the desire to be positive, encouraging, and supportive of others in their abilities and attempts to reach their goals. It is important to give back to the community something of ourselves when we are successful. By contributing economically to an historic neighborhood, I have made a tangible contribution. By committing myself to community volunteer efforts towards improving conditions in my neighborhood, increasing educational opportunities for economically disadvantaged youth, and preserving local historical atmosphere, I have returned part of my spirit to my community. Success means sharing one's own success with others, making it possible for many more than one to be successful.

Caroline Laudig graduated from Indiana Vocational Technical College in 1986 with a degree in computer programming. She currently is employed as a computer programmer/analyst in Indianapolis, IN. As a computer programmer, she has used her computer skills to teach pre-employment skills to economically disadvantaged youths at a local alternative school. She is cataloguing houses of a particular architectural style in hopes of inspiring others to preserve the historic quality of their neighborhoods.

Rebecca Topletz Lindsay
Eastfield College, TX

The time I spent at Eastfield College presented me with many opportunities for growth, both academically and personally. I had not been in a classroom since the second grade because of illness, and attending Eastfield gave me my first opportunity to interact and socialize with my peers. This interaction led me to be one of the founding members of the Not Psyched Out Club, a club for both able-bodied and disabled students, advocating access and equality for the disabled students. Learning how to advocate for others as well as myself was my first experience in leadership of my peers.

With all my education provided at home, I had not experienced any public upper-level education and had no idea what to expect. Eastfield presented me many tremendous challenges, my first classroom experience in over 10 years, and the opportunity to learn to control my time, studies, class projects, and social activities.

My instructors were quick to respond to my needs, desires, and fears with concerned and serious assistance. The campus administration was determined to assist in the development of an accessible campus that would afford an opportunity for me and other disabled students to obtain a quality start in our college education. With the support of the college, I soon overcame my fears and began to develop a determination and confidence to continue.

Since leaving Eastfield, I have obtained my master's degree and have almost finished my doctorate I married a man I met at Eastfield and now have a six-year-old daughter. My husband and I have our own business. Advocacy for the disabled is still a prime concern of mine, and I am active in the field at the local, national, and international level. None of this would have been possible without the support and friendships I obtained while attending Eastfield College.

Community, technical, and junior colleges offer an opportunity for a student to obtain an education from instructors who are willing to interact personally with their students. No one is just a number in a class. The classes are smaller, and the facilities to assist students in developing their academic strengths are greater.

It has been almost 15 years since I left Eastfield as a student, but each time I drop by for a visit I am always reassured to see that the values the school started with are still there. I could never recommend Eastfield or any of its sister colleges enough. Thank you, Eastfield!

> *Rebecca Topletz Lindsay graduated from Eastfield College, a part of the Dallas County Community College District, in 1975 with a degree in English/psychology and lives in Dallas. She is an independent microcomputer consultant noted as a vocal advocate for the disabled at local, national, and international levels.*

Sigrid V. Messemer
Valencia Community College, FL

I used to think community colleges were for those students who were a little interested in further education after high school, but not much—just enough to get a better job than a high school diploma affords.

Education was serious business for me. I was the single parent of three active youngsters and trying to finish an industrial engineering degree at night while I worked full-time during the day. It had already taken nine years from the first college class after divorce to get to this point, equivalent to junior status.

By this time, I realized that I was more interested in construction than industrial engineering. My job had more involvement with reading blueprints, estimating materials, and planning schedules. I was in between design engineering and building construction and needed to know more about the latter—especially since, as a woman, everybody knew I couldn't possibly understand construction!

For two years, I attended Valencia Community College in Orlando, FL. The associate in science degree that I earned in architectural and building construction technology allowed me to work on a par with, and sometimes better than, my male peers who had actually worked in construction labor for their educations.

It took both the class instruction and experience in the field to fully understand my work. Having the accreditation of a degree not only added to my professional acceptance, but boosted my self-confidence on the job sites.

After graduation from Valencia, I returned to the University of Central Florida and completed the ABET-accredited bachelor of science degree in engineering technology, with the operations option (as in industrial operations). By the time I graduated there, I had spent 14 years in educational pursuit!

Somehow, between work and study, I remarried, added two step-teens to my three teenagers, and settled on safety as a career path. It took even more study and certifying exams to be able to put "CSP" (Certified Safety Professional) behind my name.

For the past three years, I have been an independent safety consultant and trainer in my own company, Commercial Safety Consultants, Inc. Much of my safety work is with construction companies on the job sites. Do I still think community colleges are for the "Just a little bit, please!" students? Not at all! Those courses at Valencia added an irreplaceable dimension to my professional life and a special confidence to my well-being.

Sigrid Vigonne Messemer graduated from the West Campus of Valencia Community College in 1980 with a degree in architectural and building construction technology and now lives in Winter Park, FL. A safety consultant and trainer, she is also president and owner of her own company. She is past president of the Central Florida Chapter of American Society of Safety Engineers and recipient of the Albert G. Mowson Award for achievement.

Lane Rushing
Stanly Community College, NC

In the summer of 1978 I was at a crossroads in my life. I was a 25-year-old with a high school diploma and no career direction. In the years since high school, I had held several jobs ranging from textiles to mass transit to insurance salesman, but still something was missing. It was about this time that I heard of Stanly Community College. Sure I had known that the college was in the county seat where I lived, but I never really considered attending. While reading the local newspaper one evening, I noticed an ad for the college. The ad mentioned a new program of study being offered for the first time anywhere in the state. How was I to know that that ad and this new program would change my life forever?

Beginning in the fall of 1978, I enrolled at SCC in its biomedical equipment technology program. Here I was, a man in my mid-20s attending classes with students just out of high school. It took several weeks for me to remaster the art of studying, but soon that was just one of the small pleasures that I was to encounter in the two years I spent at SCC.

In addition to my studies, I became involved with the Student Government Association and was elected vice president my first year and president my second year. By having the opportunity to interact with fellow students, faculty, and staff, I gained self-confidence and improved my communication skills, which were priceless in my pursuit of my new career. These experiences expanded to the state level when I was elected state vice president of the North Carolina Community College Student Government Association. This was a combined group of SGA Officers from the other 57 community colleges and technical institutes in the state. In this position I was able to experience leadership qualities that a short 12 months earlier I had no idea I possessed.

Through these experiences, I left SCC a more self-confident individual. As I entered the business world, I was prepared to accept any challenge that came my way. My studies and experiences prepared me for the health care field, in which I deal with physicians, nurses, and health care administrators.

I also have returned to SCC and tried to repay the college for the quality education that I received. This has consisted of serving on the Biomedical Equipment Technology Advisory Board and reorganizing the local alumni association. I will always be thankful for and proud of the degree I received from Stanly Community College, which took me from a crossroads in my life on to my road to success as a field service representative for Physio-Control Corporation.

Lane Rushing graduated from Stanly Community College in 1980 with a degree in biomedical equipment technology and now lives in Charlotte, NC. He is a field service representative for Physio-Control Corporation—Technical Services in Reston, VA. He was active in organizing the State Community College Alumni Association in North Carolina and serves on its board of directors.

Eugene Sebastian
Lees College, KY

I graduated from high school in the spring of 1942. I wasn't sure of just what to do. My brother had been a student at Lees College during the 1941-42 school year . He and I went to Michigan to work for the summer. We saved enough money to enroll at Lees in the fall of 1942. We could stay at home and commute to Lees.

Lees was known as a good college to earn a teaching certificate. Most of the teachers in Breathitt County and the surrounding counties had attended Lees.

That year at Lees meant so much to me. Lees gave me a start to a great future. I began my teaching career in the fall of 1943. While attending Lees, I met Vada Deaton, who also was planning to be a teacher. She became my wife in October 1943. We both have had many years of success in education.

There is no way to measure the good Lees College has done for this area. You just have to appreciate that Lees was there.

Eugene Sebastian graduated from Lees College in 1946 with a degree in education and now lives in Jackson, KY. He owns Sebastian Farms and also works as a part-time instructor at Lees College.

Philip Vaden
Pellissippi State Technical Community College, TN

I had made average grades in high school. My grades were average because the subject material was presented in a way that, in my opinion, neither motivated nor encouraged me to do better. I always knew I could do better than average work, but the challenge was not there, and I only gave enough effort to maintain decent grades.

I had majored in math in high school and was encouraged by relatives to enter one of the engineering fields when I went to college. Upon entering a major university in the fall of 1980, I had high expectations of what I thought college should be like. My illusion of a hands-on learning environment quickly dissipated. The art of educating had been reduced to viewing videotapes in a standing-room-only auditorium. I found it impossible to concentrate or focus my attention on the material at hand, and I soon lost interest in my subjects. My grades fell, and after two quarters I was academically dismissed from the university.

Turning my attention to a career, I took a full-time position at a machine shop in which I had worked weekends and summers during the last few years of high school. Five years and two jobs later, I realized that technology was taking the manufacturing processes into a realm that would soon need fewer general machinists and more highly trained technicians who could program and set up computer numerical controlled manufacturing equipment. The thought of being phased out of a job is a very sobering one. I accepted the fact that returning to college would be necessary.

I applied to and was accepted by Pellissippi State Technical Community College, and immediately upon entering the college my opinions of higher education took a turn for the better. Gone were the TV sets and crowded auditoriums. In their place I found counselors and teachers—people who took an interest in me personally and challenged me to give college a chance. Pellissippi State provided me with the means to an end, which I have now used to secure a solid career in the manufacturing industry. I will be forever thankful to all those at the college for their commitment to keeping the education process in touch with the needs of both industry and the student.

Philip Vaden graduated from Pellissippi State Technical Community College in 1989 with a degree in mechanical engineering technology and now lives in Knoxville, TN. He is an SPC manager/CNC Manager for Tennessee Tool and Engineering, Inc. He volunteers his time to assist students in Pellissippi State's mechanical engineering technology laboratory.

Mildred Crowder Westerdale
Trident Technical College, SC

It was never my intention to attend Trident Technical College. The decision to attend was based on necessity, but it was probably the best decision I have ever made. My high school class ranking and membership in the National Honor Society warranted a partial scholarship to Morris Brown College; however, my family was unable to provide the balance of funds needed. The guidance counselor from my high school advised me to go to Trident and see what was offered. I did—and the rest is history. Two years later I graduated from the chemical engineering technology program and achieved my personal goal of receiving a quality education. In the 23 years since graduation, I have held many positions and accomplished many things. I served on an advisory board for the college, was instrumental in establishing the college's alumni association, and served as one of that association's presidents.

I am currently employed as a nuclear engineering technician in the Nuclear Quality Control Office at the Naval Shipyard. Any success achieved by me must be attributed to Trident. The school opened its doors with only one goal—to provide the Lowcountry of South Carolina with trained, competent individuals who could contribute immediately to the work force. Trident has been extremely successful in attaining this goal. The college has continued to grow and provide a quality education at an affordable price. The success of any person is directly related to his or her educational background, and I received mine at Trident Technical College.

Mildred Crowder Westerdale graduated from Trident Technical College in 1966 with a degree in chemical engineering technology and now lives in Hanahan, SC. Mildred helped establish the Trident Technical College Alumni Association, served as alumni association president in 1984-1985, and was Charleston County Trustee for the Alumni Association in 1982-1983. Her current occupation is nuclear engineering technician for the Charleston Naval Shipyard.

Pamela A. Wilkins
Tompkins-Cortland Community College, NY

Fifteen years ago, my high school guidance counselor discouraged me from pursuing my only dream: being a veterinarian. I was told, "That's something women just don't do." I was told to get a degree in English, then I could teach...and if my husband ever left me, I could support myself. Unfortunately, I bought that advice—hook, line, and sinker.

I enrolled at a small, private liberal arts college in northern New York state, and after just one semester I dropped out and returned home to work with horses. For nearly three years, I rode, trained, and gave lessons in Cortland and Ithaca, but it was a hard way to earn a living and the money wasn't very good.

That's when I decided to follow my heart. In the fall of 1978, I enrolled at Tompkins-Cortland Community College, which for me was a big step. Like many older students, I was scared about returning to school after a few years in the working world.

But once I was there, it was a very good experience for me. I gained a lot of confidence and found that I could do the work. I ran into people who took an interest in me and encouraged me to try. Even during difficult times in my personal and family life, I found my professors willing to help or lend a hand. The personal attention they gave me made all the difference between failure (again!) and success.

I majored in biology, with an emphasis in animal anatomy and physiology, and took most of my core courses at Tompkins-Cortland, including biology, chemistry, organic chemistry, calculus, and physics. And when I graduated, I transferred all of my credits to Cornell University, where I earned my bachelor's degree, followed by my DVM and, later, a master's degree in surgery. My academic base from Tompkins-Cortland was excellent. I had no trouble understanding the material in the advanced classes at Cornell.

In 1987 I established an equine neonatal intensive care unit at Cornell. I wrote grants for equipment, organized a course for students, rallied 100 volunteers, trained them, and finally won the administration's support for a staff position. I know that without the confidence I gained at Tompkins-Cortland, I never would have attempted such an extraordinary project.

Today, I am one of three veterinarians composing a firm that serves nearly 1,400 clients near Albany. We provide care ranging from nutritional counseling to emergency surgery. Many of my days are spent driving to and from house calls, or stable calls in this case.

For me, it's more than a job. It's my lifelong ambition. Tompkins-Cortland was a turning point in my life. It gave me the confidence to go for my greatest dream.

Pamela A. Wilkins graduated from Tompkins-Cortland Community College in 1980 with a degree in biology and now lives in Slingerlands, NY. She is currently a veterinarian and her extensive research and writing on equine colic has been published regularly. She has also been a guest presenter to many lay and professional groups, including the International Society of Veterinary Perinatoligists and the American College of Veterinary Surgeons.

Michael Wilson
Highland Community College, IL

My father, mother, and older sisters all attended the University of Illinois several miles from our farm home in northern Illinois. I expected to do the same, but during my senior year in high school, personal circumstances made it impossible. My mother was terminally ill, and it was necessary for me to stay near the farm.

I decided to apply at Highland Community College, majoring in agriculture. I had no idea what college life would be like as a commuter student. Coming from a small high school—graduating with fewer than 30 people in my class—I wondered if I could handle college life. I didn't have a clue what I wanted to do with my life at that time.

During my time at HCC, I discovered a lot of things about myself. I soon learned that I had many opportunities to get involved in activities at HCC. I participated in speech tournaments, jazz band, and student senate, and was elected student president my sophomore year. I began to realize that these kinds of opportunities would probably not have been possible at a large university.

The smaller class sizes at HCC made interacting with professors easy. And they probably improved my chances at surviving the tough freshman core classes such as chemistry.

I really learned how to communicate with people when I attended HCC. I think that's because the professors at the college really took a genuine interest in me, not just as a student, but as a person. Even today, after 10 years, I keep in touch with some of the staff members. I consider at least one HCC teacher my mentor, a person who had a definite impact on my career choices. He not only taught me to speak in public, but to think and write with clarity and honesty.

My relationships and experiences at HCC gave me good lessons for life: not only how to communicate, but also how to work with others, how to be fair, and how to think creatively.

It was at this point that I was able to focus my goals and career ambitions. After HCC I transferred to the University of Illinois and obtained my bachelor of science degree in agricultural journalism. When people ask me about my experience at a "junior" college, I tell them I would not have changed a thing about my college career. I speak of HCC with pride because it truly was a great experience.

Michael Wilson graduated from Highland Community College in 1979 with a degree in agriculture. He is currently the managing editor of Prairie Farmer *Magazine. As a professional journalist, He has devoted himself to maintaining close contact with what is happening in the field of agriculture. He is dedicated to keeping the public informed of the latest agricultural developments.*

Ken Zabielski, Sr.
McHenry County College, IL

Fourteen years ago, an opportunity arose for me to selectively change careers. Guidance and direction was offered by McHenry County College's faculty, and it has been one of the most important journeys that I have ever embarked on. The established Veterans Student Adviser assured me that nothing ever exceeds success like success itself. Reluctantly, I gave up moving 14 tons of beverages a day. They were packaged in glass bottles, and an idea of developing a replacement lightweight (plastic) bottle caught my interest.

The staff at MCC launched me into a totally new galaxy of learning. Communications was emphasized, as were the analytical skills. The ability to question and disclose information became ingrained into our shared prescribed curriculum. The renovated Information Resource Lab allowed me to expand and reach literally across the world. The scientific journals fueled a somewhat inquisitive mind. As my credits accrued, the real test came. I entered the realm of the plastics industry.

Initially, I was a silent worker and observer. I received my associate degree in plastic technology in August 1977 and then became a somewhat vocal analytical technician. Achieving the highest technical slot did not satisfy me from a knowledge standpoint. Engineering was at that time a mere dream. Still, with the associate degree I realized that dream and became a research and development engineer.

Just last semester, I returned to MCC for a computer course. As I entered those halls, my eyes swelled and teared up. I took a deep breath and then realized what had transpired. I mumbled to myself "Twelve years and 21 patents later, I come back, still with my AAS degree, and still on a quest for knowledge. Not a bad life's experience after all."

Currently, I am attending the National College of Education. Who knows where a bachelor's degree will lead me?

I offer the following for deep consideration. First, allow yourself to reach and take advantage of your community college. Second, never ever stop learning. Geographically, a community college is very close to your home. Educationally, the curricula are custom-tailored. Impact-wise, I can only say that the last patent issued to me (February of 1988) is the subject of the 1990 International Netlon Gold Medal. It is awarded for innovative plastics and rubber processes. Only one is bestowed every five years.

Thanks, MCC. You now have my children. If you can be kind enough to just lightly brush that learning wand across their brows, it will be ever appreciated. If you could touch any other student as you embraced me, the world will indeed prosper. Above all else, MCC, remember that..."we are the world!"

> *Ken Zabielski, Sr. graduated from McHenry County College, IL in 1977 with a degree in plastics technology. He is area engineer for Quantum Performance Films, Quantum Chemical Corporation in Streamwood, IL.*

CHAPTER FIVE

Entertainment, Media, and Sports

Dennis Anderson
Los Angeles Pierce College, CA

When I signed up for the fall semester at Los Angeles Pierce College in 1975, my duffel bag was still crammed with military whatnot from a three-year infantry hitch in West Germany. I had gone into the Army because I didn't know how to study after high school and couldn't keep my mind on class. I ended up at Pierce College because the campus was pretty and because during leave I had visited and the counselors promised a warm welcome for vets, which turned out to be true.

In the passage between high school and the military, I had cultivated a phobia of universities. They seemed distant, forbidding, and lofty. I felt sure that if I got out of the Army and tried to plunge into academia at such a place I would surely sink. Thus, the community college system became my first stop, a place to wade in and get my feet wet without drowning.

My first stop was the journalism department, because I wanted to write and I hoped someone there could show me how. I met a half-dozen of the finest teachers I will ever know. They combined the demanding aspects of military instructors with the intellectual horsepower of true mentors. Most importantly, they cared, passionately. Later, at the two universities I attended in the course of finishing my studies, I encountered two or three memorable professors. From Pierce College I have a half-dozen such personal and professional relationships. That is a startling equation. The teachers I knew at Pierce were in every way the equal of their university peers. In some ways, they exceeded their university brethren. Free of the pressure of "publish or perish" and less involved in faculty politics, my Pierce teachers invested their time where it belonged, in the students. This is not fluff or empty homage; it's a case study, and I think there are a lot of similar stories out there.

When I was "adopted" by the journalism faculty, with my duffel bag still stuffed, if I was a diamond, I was certainly rough. Those guys know how to romance a stone, to get it shiny and buff it up to a fine luster. They did such a good job that in 1977 I became editor of the student newspaper, *Roundup*.

When I entered college, I had neither money, grades, nor family connection to pave the way for success. But even without these blessings, a motivated community college grad can put together the pieces that combine to make up success in a very competitive America. This is important and should be noted at all levels of government and within the communities that support the community college system.

> *Dennis Anderson graduated from Los Angeles Pierce College in 1977 with a degree in journalism and now lives in Antelope Valley, CA. He is a novelist and night editor for Associated Press in Los Angeles. He is the author of the novel* Target Stealth, *published by Warner Books. He is also a former paratrooper and recognition specialist for NATO.*

Ed Arnold

Texarkana College, TX, and Rancho Santiago College, CA

When I was a youngster, the subject of college was never discussed. It could have been because no one in my family had earned a degree. Most had graduated from high school, but that meant...it was time to work 40 hours each week at minimum wage. A college education was for the privileged.

The change came for me when a junior/community college opened in my hometown. Though there was still no discussion about college, and my high school teachers rarely broached the topic, I became very aware of the new school. However, my bubble burst because of my need to work. That's when it hit me. I could work <u>and</u> attend Texarkana College. I did, and I joined the Platoon Leaders Program of the U.S. Marine Corps, which would earn me extra money. When I graduated, I would serve the Marine Corps as an officer.

Ah, the dreams of a kid. Personal and financial problems hit during my first semester of college, and I had to go on active duty in the Marines. But I had discovered college, and the burning desire remained with me during my years in the Marines. There I was encouraged to better myself through courses. I also had a chance to participate in organized sports.

Our coach scheduled football games against local community colleges so we could get exposure to the schools and have the coaches see us play. He knew those coaches would be interested in recruiting players who could help their programs. One of those coaches was Homer Beatty from Santa Ana College, since renamed Rancho Santiago College. I was lucky enough to be asked to attend. I know I was not an outstanding player, so I was very fortunate.

A community college again played an important part in my life. My professors at RSC were encouraging and helpful as I began my formal schooling once more. After a year I again had to drop out because of financial problems, but the desire was still there. I returned a couple of years later, and the rest is history. I earned my AA from Rancho Santiago College, graduated from California State University, Long Beach with a bachelor's degree, and was guided from radio to television, an occupation I still hold.

And...I became the first member of my family to earn a college degree!

Ed Arnold graduated from Rancho Santiago College in 1966 with a degree in communications. He is currently working as a sports reporter/anchor at KTLA Television in Los Angeles and resides in Fountain Valley, CA. He is recognized by thousands for his work in public and commercial radio and television. Less known but equally important are his significant contributions as president, board member, and active participant with a number of organizations representing such interests as multiple sclerosis, Special Olympics, cystic fibrosis, March of Dimes, Boy Scouts, Jaycees, Rotary Club and countless others. He has earned many honors, both for his broadcasting and for his volunteer efforts.

James G. Baldridge
Sinclair Community College, OH

Community colleges are America's opportunity makers. They offer a chance at higher education—and success in life—to people who, for one reason or another, might have missed it.

People like me. As a teenager in Lima, OH, I was bright enough but not particularly interested in school. Broadcasting was my passion. After high school I found a job at a local radio station. I started college because my parents insisted, but my heart wasn't in it. Within a few months I dropped out. My local draft board noticed and I soon found myself in the Army working for Armed Forces Radio and Television. My stay in the military matured me. By the time the Army was through with me, I was ready for a second chance at an education.

Fortunately, Sinclair Community College was ready for me. Because Sinclair offered night and weekend classes, I was able to fit school around my erratic work schedule as a junior news writer at a local television station. It was thrilling. Liberal arts studies opened doors to culture and knowledge. Teachers with "real world" experience brought special insight. This time I found myself not only doing well in school, but also enjoying it.

Even more important than the new things I learned—and the associate's degree I received—was the fact that Sinclair helped me to organize my mind and better understand the things I thought I already knew. That, I believe, is true human growth.

Success at Sinclair led to success elsewhere. I earned a bachelor's degree at a nearby state university and made a start at graduate work. My career has been fulfilling. As a television journalist I have traveled to Europe, Central America, and all over the United States covering everything from earthquakes and airplane crashes to crime and politics. I am now senior anchor at WHIO-TV in Dayton, OH.

Sinclair Community College is where I turned the corner. I try to repay the community by visiting high school and college classrooms to share my experiences and stress the importance of higher education. I have been able to use my prominence in local television to help area charities.

James G. Baldridge graduated from Sinclair Community College in 1973 with a degree in liberal arts and sciences. He is currently employed as a television news anchor and resides in Centerville, OH. He is a frequent speaker at local schools on the importance of higher education. His community service includes involvement with numerous fund raising activities for local charities, membership on the Cystic Fibrosis Board, and past board membership in the American Cancer Society. In 1989 he was selected as a distinguished alumnus of Shawnee High School in Lima, OH. He has also received the Ohio Foreign Language Award, "Friend of Foreign Languages," for stimulating young people to learn second and third languages.

One of my most satisfying moments came recently, when I reported that voters had approved a new long-term tax to keep Sinclair Community College doing for others what it had done for me. My community had chosen to keep one of its most vital resources thriving...AND offer all of its citizens the opportunity for success.

Marilyn Benner
Florence-Darlington Technical College, SC

How did FDTC prepare me to survive the daily needs of show business? My instructors wouldn't settle for anything less than my very best performance...no matter what. The "plot" was very specific ... be prepared for anything, anticipate potential complications, and above all be a professional at all times. Spot problems before it's too late. Gather information. Don't be afraid to make decisions. Set realistic goals. These lessons have proven invaluable for both my business and my patients.

On many occasions, I reflect back on my days at Florence-Darlington Technical College, where I obtained my associate degree in nursing. I'm very thankful that they found value in me to accept me from a list of 300 waiting applicants. I wanted to be a nurse more than anything, and I convinced them that it was not an overnight decision. I feel that I received the best nursing education possible at Florence-Darlington Technical College.

It was a very comprehensive and fast-paced program, and I spent the two years studying, studying, studying. The competition was tough, and so were my instructors. They knew they were preparing us to take lives into our hands every day. As I look back, I feel that I came out better prepared clinically for my job than most bachelor's-degree-prepared nurses. Why? Because we spent much more time in the clinical area... performing procedures in real patient settings rather than sitting in a classroom reading about them.

"How to turn an emergency room drama into a simple operation." This is the theme of my company, Benner Medical Productions. We provide medical-technical consultation to the movie and television industry. Only three years old, my company is now working on its 15th major movie, in addition to the daily productions of "Guiding Light" and "Another World." The mission of Benner Medical Productions is to ensure that our audience of millions of viewers receive correct medical information through the programs they watch for entertainment. This means we get involved in correcting dialogue as well as creating medical stories to fit the original plots. Not only do we act our parts in front of the camera, we also equip the set, coach actors on medical procedures and pronounciation of terms, and coordinate and provide medical emergency teams as backup for stunts.

I came to New York to study AIDS at The Mount Sinai Hospital... and ended up doing the first AIDS storyline on daytime television.

Even with my demanding schedule in running the business, I am still on staff at one of New York's "craziest" trauma units. Thanks to Florence-Darlington Technical College, nursing is still the focus of my life.

Marilyn Benner graduated from Florence-Darlington Technical College in 1976 with a degree in nursing and is now living in New York City. She is president of Benner Medical Productions, Inc. As an actress, writer, producer, medical consultant, and certified health professional, she is in a position to ensure accurate portrayal of medical situations in the entertainment industry.

Larry Downing
Los Angeles Pierce College, CA

Since I am continually exposed to the glamour, excitement, and sophistication surrounding Washington, D.C. and other capital cities around the world, one might assume that I have forgotten my much quieter college days. But this is not the case. I hold warm and lasting memories of my tenure as photographer, reporter, and then editor of the Pierce College *Roundup*. I think I owe everything I've achieved to the instructors at Pierce.

My rigorous training at Pierce has helped me chalk up some prizes, including first and third place in the National Press Photographers Association Picture of the Year-Generic News for coverage of President Carter in 1979. I also received first place in the National Headliners Award for Excellence in Magazine Photography in 1984 and third place in the World Press Photo Spot News Contest for Reagan and Gorbachev coverage in Geneva, Switzerland, in 1985. I won the NPPA Award for Excellence in Photography for coverage of the White House in 1986, and I have garnered numerous first, second, and third place awards in the White House News Photographers Association Contest since 1979.

The individualized instruction I got at Pierce from teachers devoted day and night and on weekends to helping students learn the craft of journalism and photography enabled me to survive and then thrive on the fierce competition I face every working day. A fact of life for Washington photographers is that only the quickest, most capable, and most competitive last long. This is because Washington editors are among the most demanding in the world on a daily basis.

The public relates to photography the way it does to many glamour professions, by seeing the photo images and having no idea about the intense concentration and hard work that go into making those pictures. This is the kind of discipline I learned from my very demanding instructors at Pierce. Photo opportunities at the White House are not as glamorous as you might think. It is almost a contact sport. You don't have lots of time. You don't have time to second-guess or think very much. But covering Reagan since before the 1980 presidential campaign gave me an edge. I could tell by the position of his fingers what he was going to do.

I enjoy competition. That is something I learned about myself at Pierce, when I turned my hobby of photography into a profession. It happened by accident. I had originally enrolled at Pierce College to become a lawyer. To sharpen my writing skills, I took journalism courses, and the classes and the professors who taught them forever changed my life.

Larry Downing graduated from Los Angeles Pierce College in 1975 with a degree in journalism and now lives in Washington, D.C. He is a photographer for Newsweek *and a member of the Washington, D.C. Press Corps.*

Daniel L. Dreisbach
Greenville Technical College, SC

If there is one word that characterizes my educational background, it is "unconventional." I was born and raised in West Africa on the frontier of the vast Sahara Desert, the son of medical missionaries. I received my early formal education at home and at boarding schools in Africa, Europe, and America. My childhood experiences in Africa and the example of Christian and humanitarian commitment set by my parents were the most significant factors in shaping my values and ordering my goals.

Following graduation from high school, I set out for Africa as a volunteer laboring among the drought-stricken tribes of the Sahel. I returned to the United States in 1978 eager for further education, but my choice of colleges was limited by financial constraints and family responsibilities.

Greenville Technical College offered the opportunity I desired. Tech's two-year associate in arts degree in the humanities had an established reputation for quality and affordability. Most importantly, Tech offered essential courses morning, afternoon, and evening, which enabled me to schedule classes around a full-time job as an orderly in a local hospital. At Tech I not only obtained transfer credits in foundational courses, but I also explored diverse academic interests through the rich and varied curriculum.

I completed the associate degree in 1980 and transferred to the University of South Carolina at Spartanburg, where I earned a bachelor of arts degree in government. In my senior year I was awarded a Rhodes Scholarship that provided four years of study at Oxford University, England, where in 1985 I completed a doctor of philosophy degree, specializing in constitutional law. This was followed by a juris doctor degree from the University of Virginia and a rewarding term as a judicial clerk in the United States Court of Appeals for the Fourth Circuit.

I am now a writer completing my second book on American constitutional law. Greenville Tech provided the launching pad I needed for a satisfying and rewarding career as a writer, lawyer, and academic. It is always with great pride and satisfaction that I recount that my path through higher education commenced at Greenville Technical College.

> *Daniel L. Dreisbach graduated from Greenville Technical College in 1980 with an AA degree. He currently is a lawyer and writer and resides in Charlottesville, VA. He is one of the few community college graduates who have been named a Rhodes Scholar. He has rapidly become a noted author and lecturer. His first book won critical acclaim for its treatment of the subject of religious freedom and church-state relations in America. He is regarded as a leading authority on constitutional law and religious liberties.*

Faire R. Edwards
Community College of Vermont , VT

Some years ago, I was still adjusting to widowhood and looking at ways to support myself at an interesting and useful job. I also was recovering from an ulcer and had gone for counseling. My counselor suggested that I might find it helpful to complete the requirements for an associate degree at the Community College of Vermont.

I signed up and won Clo Pitkin for my course counselor. It was a great experience. Even when Clo was working with someone with a very different background or experience, I found the group action helped me get my brains together and discover how much I had really learned after I had gone to work, married, reared a family of six, and now faced the rest of my life with at least a decade of being a one-person household doing the kind of work, for pay or for satisfaction, that I would find most productive.

Now I know that, from time to time, I need to go back to school and put my theories and current duties together and keep them in line toward specific goals, being an activist toward making democracy work for the good of most of the society where I can reach it.

Facing the last part of one's life is a challenge in these days of modern technology. Born into a world of oil lamps, trolley cars, and very few medications that prevent illness or deal directly with bacteria or viruses, I have seen wonders. So have my contemporaries who have watched the democratic revolution make historic changes in our national lifestyles.

When you live in a fading body, you also have to find ways to keep both mind and anatomy well enough adjusted so that you can think and act consistently. You have to be able to draw upon old skills, developed long ago. Your high school debating experience can be refurbished to assist you in speaking ad lib before a legislative committee. There is an excitement in seeing what you can still do, and in finding victories to cherish.

Now I'm enrolled in the year-long gerontology class at the University of Vermont's continuing education program, along with other students barely half my age. Because I am no longer employed regularly, I can speak out without enraging an employer. Just recently I have found this will be necessary, because I was the only person in our class who is in a position to use facts we have been taught. The need for advocacy became vividly clear when I realized that we were getting information about Alzheimer's Disease which the State Health Policy Council had not included in its State Health Plan. I hold a couple of offices that will give me a good base from which to speak, and my professor is going to insist on factual and workable efforts to work on the project I found for myself in the classroom. Many older people possess great memories and skills and observations. We also need to keep our thinking in up-to-date form. I hope that I can be of enough value to justify the free tuition available to senior citizens from either the Community College of Vermont or the state colleges in Vermont.

Faire R. Edwards graduated from the Community College of Vermont in 1974 with a degree in communication and now lives in Montpelier, VT. She is a radio commentator, freelance writer, and activist.

Frances Garmon
Temple Junior College, TX

The fact that I was living on a farm in the small community of Willow Grove only 15 miles from Temple, TX, made Temple Junior College an attractive choice for continuing my education when I graduated from Moody High School.

The professors at Temple Junior College were, and I believe still are, truly the best-kept secret around. The educational background I received was exceptional. Almost everyone can remember a special teacher. There were many of these on the staff at TJC during my time there.

I always had the feeling—as other students did—that I was individually valued. The administration and board of regents at Temple had to be great judges of what it takes to be a teacher who could challenge students fresh from high school, as well as the older adults attending the college.

My weekends were spent traveling with an independent basketball team coached by Imogene Sodek. The opportunity to play basketball with outstanding players from central Texas just added to my career preparation and my pleasure.

Following graduation from Temple Junior College, I attended Mary Hardin-Baylor University were I received a bachelor's degree, and then completed my master's degree at Baylor University. I have since completed postgraduate work at other universities.

Temple Junior College was the first step in an exciting teaching and coaching career. Seventeen years of my teaching and coaching career were spent at TJC. I know first-hand the influence a junior college can have. I may have been a confused country girl from Willow Grove, but at TJC I was treated like a special person.

TJC allowed me to grow as a student and a teacher. The trust and understanding of the TJC administration gave me the opportunity to develop a women's basketball program respected throughout the nation. I was also allowed to participate in the development of a growing international basketball program in the USA.

I am grateful for the continual support given me by everyone at TJC in reaching goals most only dream or read about, such as coaching a national championship team, a World University Championship team, a Pan American Championship team, and being the coach of the first team to defeat the Soviet Union in a communist-block country. These are just a few of the exciting moments in sports I have experienced.

Believing in people and allowing them to grow, no matter how big or small their goals, makes Temple Junior College number one in my book.

Frances Garmon graduated from Temple Junior College in 1959 with an associate in arts degree. She is currently the womens basketball coach at Texas Christian University and resides in Fort Worth, TX. She was a pioneer in the development of womens basketball at the collegiate level and was a charter member and past president of the Womens Basketball Coaches Association. She was the first recipient of the prestigious Wade Trophy Coach of the Year Award. Through the years her squads have achieved success in the classroom as well as on the court. During her span as coach she has seen 150 of 156 student athletes earn at least a bachelor's degree.

M. Gasby Greely
Wayne County Community College, MI

In 1969 I was divorced and the mother of a three-year-old daughter. As head of the household, I worked in a spark plug factory to support us. While I wasn't sure about what I wanted to do in life, I knew that inspecting and making spark plugs wasn't it. Because I hadn't attended school since graduating from high school four years earlier, I felt insecure about tackling college entrance applications and exams. Wayne County Community College had an open-door policy, and there was a satellite campus near my home. This was all I needed to get started.

WCCC gave me the confidence to achieve bigger and better things. The teachers and the atmosphere inspired me to look at my goals from a different perspective; I stopped putting a ceiling on what I could accomplish.

After graduating from WCCC with an associate of arts degree, I enrolled at Wayne State University and majored in psychology. After graduating, my professional career was launched: I worked as assistant to the vice president of community affairs at Detroit radio station WGPR-FM, went on to public relations at United Community Services of Metropolitan Detroit, and then spearheaded community outreach activities in three states for the U.S. Census Bureau. Later I became the Bureau's national spokesperson, appearing on radio and television shows across the country.

When the 1980 Census was over, I entered Harvard University's Kennedy School of Government. I graduated with a master's of public administration specializing in communications management, then moved to New York to become marketing director for WNET-TV, the nation's largest PBS station. But another broadcasting career was waiting for me. Today I have my own talk show on New York's WNYC-TV, hold several correspondent positions for Fox Television and Black Entertainment Television, and have my own business.

For the past five years I have volunteered to counsel and motivate female inmates at Bayview Correctional Facility in New York City. Their need to see hope in the face of despair inspires me to give of my time.

M. Gasby Greely graduated from Wayne County Community College in 1971 with an associate of arts degree. Today she is a television reporter working and living in New York City. Before she attended WCCC, she worked in a spark plug factory to support herself and her young daughter. Today she is a graduate of Harvard University, a television journalist, an adjunct professor at two urban colleges, and president of her own media/communication company. She has won numerous awards for her work in communications and was chosen Volunteer of the Year for her work counseling female inmates at Bayview Correctional Facility in New York City.

As a result of encouragement and training I received at WCCC, I am keenly interested in giving something back. This interest led to an adjunct professor's position at Manhattan Community College. Today, I am an adjunct professor at Harlem's City College. My hope is that I can do for students what the teachers at WCCC did for me.

WCCC was my springboard to higher education. My accomplishments are due to the support of a loving family, an understanding daughter, my will to succeed, WCCC and, most of all, to the grace of God who is with me every step of the way.

Ted B. Hall
Isothermal Community College, NC

I wouldn't know J. J. Tarlton if I met him on the street, even though I've heard of his work on behalf of education in Rutherford County, NC, all my life. And he wouldn't know me, either. But Mr. Tarlton changed my life and I owe him undying gratitude.

Twenty-five years ago, he was chairman of a group which fought for and won a community college in Rutherford County. Four years later, Isothermal Community College was there when a boy from the mill hill needed it.

The damp basements which served as the first Isothermal Community College classrooms were gone by the time I came along. The college already had progressed to three modern buildings beside a lovely little lake in the country.

Saplings struggling to hold on when I first attended Isothermal are beautiful trees now. Buildings have spread out all over the hillside. And enrollment has grown from an initial 124 to about 500 when I was there from 1968 to 1970, to more than 2,000 today.

For me, and I'm sure for many others like me, Isothermal Community College provided the springboard I needed. Without it, I probably wouldn't have gone to college.

I was married and working in the card room at a textile plant when I enrolled at Isothermal. The state's public universities were beyond my means. I was the student for whom community colleges were designed. With their low tuition fees and convenient locations, they were affordable, even for people like me.

And with their small size and friendly style, they offered good educational opportunities. Every course I took at Isothermal transferred to the University of North Carolina at Chapel Hill. And I never once felt at UNC that I was getting better instruction there than I had received at my community college.

Teachers at Isothermal not only taught me the course material I needed before going on to a four-year school, but they also took me by the hand and showed me how to apply for loans and scholarships that would make it possible to continue my education.

Working in a textile plant is an honorable occupation. My father did it most of his life. And had I stayed in that job I would be proud of my work. But my point is that Isothermal Community College expanded my options and gave me the opportunity to pursue other careers. Without it, there would have been few choices.

Choices and opportunities are what J.J. Tarlton and others like him provided Rutherford Countians when they succeeded in establishing the community college. North Carolina's system of community colleges offers similar advantages to students all over the state.

May we never underestimate the value of community colleges in extending educational opportunities to those who otherwise might be left out.

> *Ted B. Hall graduated from Isothermal Community College in 1970 with a degree in general studies. As newspaper publisher for the* Shelby Daily Star, *his place of residence is Shelby, NC.*

Alan Bomar Jones
Sinclair Community College, OH

In the fall of 1977 I was ready for a new challenge. I had just been discharged from the United States Navy and had spent the summer relaxing and collecting unemployment. I was browsing through the newspaper in the want ads and discovered a small promotional ad for Sinclair Community College. What a great idea! I always enjoyed school, and college sounded like the challenge I needed.

That following Monday I filled out the proper forms for entry and began pondering what classes might interest me the most. While in the Navy I had the pleasure of working part-time as a movie projector operator. From that experience I took a great interest in acting, so acting classes sounded ideal.

I have to admit that I was having a blast in my acting and dancing courses. After a year into my degree, I was noticed in one of the college productions by a local television station employee. He offered me the chance to act in what was known as "Con of the Week," where I would portray either the criminal or the victim. From there I was offered jobs in commercial voice-overs and as a master of ceremonies in local fashion shows. My career took an early start. Upon graduation I moved to Seattle, WA, where I lived for one year before returning to Dayton. In Seattle I traveled as a master of ceremonies in an entertainment group, which took me to Idaho, Colorado, Hawaii, and finally Alaska. After returning to Dayton, I began teaching aerobics. I won third place in a national aerobic championship that was filmed in Hollywood. Then I hosted my own aerobic workout program, "Bodyfire," that won best new locally produced show of 1986. Following that success, I produced my own workout video, "The All-American Workout," which is presently selling well in Dayton.

I have settled in my career in radio broadcasting, hoping it will lead back to television. I am presently pursuing a bachelor of arts degree in communications. All this great success began at Sinclair Community College. Thanks to my discovery there, I have been discovered ever since.

Alan Bomar Jones graduated from Sinclair Community College in 1980 with a degree in communications. He currently works as both a radio broadcaster and an aerobics instructor and resides in Trotwood, OH. As host and choreographer for the aerobics fitness entertainment show "Bodyfire," he was honored as "producer of the year" in 1984 by the Miami Valley Cable Council. In addition to organizing "The All-American Aerobic Championship" and being a two-time regional winner and finalist in the Crystal Light National Aerobic Championship, he has produced and directed his own workout video marketed nationally.

Monte Markham
Palm Beach Community College, FL

Palm Beach Community College provided both the time and the opportunity for me to make a very successful transition from high school to college. My grades in high school were average. Like most of my peers, my efforts had been directed toward getting by and having a terrific time in the process.

I wanted very much to go to college, but the thought of actually being there was daunting. The expense of attending a university was out of the question, so with a bed at home and a good part-time job, I enrolled at PBCC.

The experiences I received enabled me to go on and compete at the highest level in university graduate and postgraduate work.

One of the best-kept secrets in the academic world is that community colleges are staffed by many first-class professors. In a university, mostly junior instructors have to take on all general requirement courses for freshmen and sophomores—the so-called "Factory Sessions." This is not to say that they aren't good at what they do, but there is an attitude of, "I should be teaching upper-level courses, and can't wait till I get there."

In contrast, community college teachers usually have you for courses during both years. But most important, their courses are structured as a more complete end in themselves—a part of the liberal arts degree. This is a unique feature of community colleges and why, I believe, excellent teachers make them their career.

But an equally important feature is the word "community." The unusual spread of age and experience among the students provides an ease, a familiarity, a sense of belonging to the college as a community. It's a place that gives you room to grow as you learn.

I've always known it was the assurance, the confidence, the hunger I gained during my years at PBCC that provided the possibilities for what has been, so far, a uniquely rewarding life. I strongly recommend the community college experience. How could I do otherwise?

Monte Markham graduated from Palm Beach Community College in 1955 with an AA degree and now lives in Malibu, CA. He is a distinguished actor on the stage, screen, and television. He is a scholar and has served as actor-in-residence at various universities. He is listed in Who's Who *and has served as a lieutenant in the U.S. Coast Guard Reserve.*

Evelyn Meine
Oakton Community College, IL

What a difference two decades make!

Oakton Community College and I grew together. Oakton from an unpretentious, industrial park campus-of-sorts to a splendid gem of a real campus—and I from a hopeless blob to a woman who has had a great time unraveling all those strings that had been holding back the person I really WAS and COULD be—and I'm not finished yet!

In 1970, I faced a challenge that nearly did me in. A bout with cancer, a 20-year marriage on the rocks, four sons still in school who needed me, and no income of my own. I knew I had to do something to survive and to keep my family together.

I don't remember exactly what led me to Oakton Community College, but before I could explain it, I was enrolling for my first legitimate classes in over 25 years.

Me! A college freshman. It was crazy! But it was wonderful. Day by day, my excitement over returning to school began to renew my faith in myself.

So...what did Oakton do for me?

- It offered me new friends, new experiences, new dreams
- It reopened doors to enlightenment that I thought had been closed to me forever
- It forced me to use my head and to become a productive human being again
- It welcomed me with loving arms and made me feel whole
- It restored my self-esteem

In 1973 I graduated. Armed with my AA degree, I was now ready for the next hurdle. What to do? Where to go? I went right back to school. I entered DePaul University, enrolling in an innovative program whereby I could earn my bachelor's degree while holding down a job.

Filled with determination and hope for the future, I applied and was accepted for a position with the Chicago Symphony Orchestra. What a wonderful late career it became. My work encompassed outreach programs for audiences of all ages, from all segments of society. The potpourri of individuals who have been a part of my life includes teachers and clowns, musicians and magicians, youngsters and oldsters, governors and mayors, congressmen and actors, writers and cartoonists. I've worked with staff and volunteers, planned parades, pushed wheelchairs, organized conferences, dabbled in media, and managed dozens of special events.

My current project is planning the centennial of the Chicago Symphony Orchestra. And when that's over (1992), I'll be ready for the next chapter. Who knows what it will be, but it all started at OCC!

Evelyn Meine graduated from Oakton Community College in 1973 with an associate of arts degree and now lives in Chicago. She has been the recipient of the following awards: Distinguished Alumni Award, DePaul University; Governor's Award for the Arts in Illinois; ICCTA Distinguished Community College Alumnus; and Who's Who of American Women.

Lisa Mendoza
Kings River Community College, CA

There is a saying that goes, "No bird soars too high if it soars with its own wings." My experience at Kings River Community College made me realize I had wings.

When I enrolled at Kings River, I was only the second person in my farm-worker family to attend college. When I registered little did I know that I had changed the direction of my life. During my stay at Kings River I learned about math, music, and myself. I left with an associate degree in liberal studies and with friends who have stayed with me until today.

After graduating from the two-year school, I transferred to the local state university. The choice was a natural one, an easy one, because my time at KRCC showed me the value of a good education.

I landed my first TV news job even before receiving my bachelor's degree in journalism. Since then I have reported, anchored, and hosted a half-hour news show. I've won some awards, but I'm most proud of a 30-minute report I did on panic anxiety attacks. Right after the report aired and for days afterward, I received calls from viewers who said they cried when they saw the report. They said all this time they thought they were crazy and had suffered their panic anxiety attacks in silence. The report led to the formation of a panic anxiety support group at a local counseling center.

Right now I'm 31 years old. I'm in TV news management and have almost 10 years of news experience under my belt. I'm ready to fly again with those wings I discovered at Kings River Community College.

Lisa Mendoza graduated from Kings River Community College in 1979 with a degree in journalism. She is currently a news producer living in Kingsburg, CA. Through her work at the local network affiliate, she has served as a very visual role model. As a Hispanic woman in a community with a large Hispanic population, she has inspired many with her ambition and dedication.

Warren Moon
West Los Angeles College, CA

When I graduated from Hamilton High School in Los Angeles, I knew that I wasn't ready for the impersonal environment of a large four-year university. I needed to be someplace where I wouldn't get lost in the crowd, where the professors would have the time to give me the individualized attention I needed instead of putting me in large classes of 200 or 300 freshmen.

At West Los Angeles College, I received that kind of special attention, and I also got a solid academic background that enabled me to transfer to the University of Washington and continue my studies there. West also gave me the opportunity to show what I could do on the football field, too, which I believe helped me to get a better scholarship when I was ready to go on to a university.

I urge all young people to go on to college and complete their studies. With a college education, you have so many more options throughout your whole life, and you can choose the career that's right for you instead of just taking what comes along.

I'll always be grateful to West for the great education I received, and for all the encouragement they gave me.

> *Warren Moon graduated from West Los Angeles College in 1973 with a degree in general studies and now lives in Houston, TX. He is quarterback for the Houston Oilers football team. He was named to the West Los Angeles College Football Hall of Fame.*

Dana Moore
Tri-County Technical College, SC

From terrified student to confident broadcaster was a trip I took several years ago, and Tri-County Technical College helped me every step of the way.

I decided in December of 1980 to re-enter college after a 12-year hiatus from school. During that time, I had been a full-time mother, a part-time hospital worker, a secretary, and had dabbled in other things before I decided what I needed was more education.

I remember well the first time I set foot in a classroom the second time around. I was scared to death. First, I didn't think my grades were good enough to get back into college. I didn't know that Tri-County Tech was an open-door institution, so I felt I had to prove myself. I began by taking just one course, freshman English, at night. I was so scared that I fully expected to receive an "F" on my first paper. Instead, I received an "A," and my self-confidence was off to a good start.

I soon found out that not only the front door was open, but also all of the office doors at Tech were open to encourage and guide me toward my career.

I majored in radio and television broadcasting but utilized just about every department at Tech and became first-name-basis friends with most professors. The students in the RTV department were like brothers and sisters. It was like being in a big family.

It took every penny I could scrape together to be able to attend college. I had a job at Tech, a part-time job at a local newspaper office, and received financial aid through a Pell grant. I was editor of the college's newspaper, *The Prism*, which was the equivalent of getting a one-quarter scholarship.

Then came the big day—graduation—and my first job. I was employed as assistant news director at WZLI in Toccoa, GA. On a whim, 18 months later, I decided to try the larger markets. My second job was where I have been for the past six years, at WFBC as a morning news anchor.

Looking back, I can see how Tri-County Tech helped me along the journey. Charlie Jordan, head of the RTV department, was like a big brother, nudging me, urging me to take chances. Charlie tells it like it is. He knows the world of broadcasting, and his program is very much like it is in the real world.

I can see how every course I had at Tech helped to prepare me for my job. For instance, I don't know how anyone makes it out in the world without a psychology course. The more you learn, the more you want to learn to keep going and to expand your knowledge.

I have been able to stay in touch with Tech. When asked to become the first president of its alumni association, I couldn't refuse. I got so much out of my experience at Tri-County, I wanted to do something in return.

> *Dana Moore graduated from Tri-County Technical College in 1982 with a degree in radio and television broadcasting and now lives in Clemson, SC. She works for WFBC Radio in Greenville, SC, as a morning news anchor.*

Ernest C. Over
Central Wyoming College, WY

Like many students entering college, I hadn't decided on a career direction coming out of high school. As a matter of fact, it wasn't until half-way through my summer vacation that I even decided to attend college. I was thinking that I'd take some time off from school and work to earn my tuition to the University of Wyoming. Instead, I enrolled at Central Wyoming College in the fall of 1969.

I decided that attending CWC would give me a chance to adjust to a campus atmosphere and to take all the basic courses I would need at a larger school. I would be close to home, and I could keep the part-time job I had at the local radio station. All the elements seemed to be there to make this decision the best one for me.

What I didn't know at the time was that my experiences at CWC established a blueprint for my successes in life. I was encouraged to be creative and to participate in a wide range of activities. With the support of my professors, I learned photography, I started a student newspaper, I became involved in student government, and I participated in theatrical productions on campus, all of which allowed me to acquire special skills that I later used to my advantage. I thrived on the campus atmosphere and expanded my radio/television curriculum to include journalism and, later, law enforcement. I made the dean's list and then the president's honor roll. I discovered that I enjoyed learning and sharing information.

Looking back at my college experience, I realize that it was at CWC that I learned the value of participation and contribution. By participating in varied activities, I began to understand what my strengths were, and I gained a much clearer career focus. I learned how to have fun, to take risks, and to not be afraid of trying something new. I gained a sense of community.

I was not only prepared to succeed in life, I was primed to make a significant contribution.

College was an enjoyable experience for me. To this day, I continue to practice the lessons learned there and integrate them into all aspects of my life.

Ernest C. Over graduated from Central Wyoming College in 1972 with an AAS degree in radio/television. He is currently serving as personal production assistant to Gene Roddenberry for the television series "Star Trek: The Next Generation." Besides work in radio as news and sports director of stations in Wyoming and Colorado, he worked as director of public relations for the Wyoming Travel Commission, promoting the state's beauty. Since moving to California to work in television, he has worked on the exhibition of the National AIDS Quilt at UCLA and with a consciousness-raising workshop.

Nick Palantzas
Massasoit Community College, MA

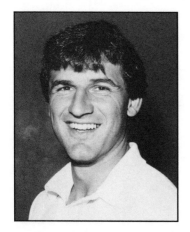

To set a goal and then accomplish it makes you feel as though you've conquered the world. That's the way I've approached my life, first as a student, and now as a coach and teacher.

Perhaps because I came to this country from Greece, staying close to my family was especially important to me. Massasoit Community College in Brockton had outstanding baseball and soccer teams, and Mike Russo, the former coach, took the time to recruit me and three others from Brockton High School.

But it was the professors at MCC who helped me, as many coaches did to keep up my academic record and my athletic involvement. As a result, I was able to transfer after two years and receive my bachelor's degree from Eckerd College in St. Petersburg, FL, in the spring of 1980, majoring in physical education. Then I came back to Massasoit to coach.

That first year, despite having only two returning players, we won the New England Championship, and we've done that six years out of the last eight. In 1986 and 1987, we came home with the National Junior College Soccer Championship, back-to-back wins. It made me want to jump for joy.

Massasoit is a good place for sports, with fine facilities, but it's a place where we keep a close watch on our athletes to make sure they get a good education as well as a place on a championship team. I try to show recruits the academic side of our school—the ratio of students to instructors and how easy it is to meet with a professor. I keep tabs on all my athletes with regular requests that their classroom teachers notify me at once of any problems that might be arising, so we can get right on it and not wait till a semester is over and wasted. In my job as academic coordinator for Massasoit's athletes, I make sure my soccer players are not just eligible to play, but are also good candidates for transfer to four-year schools. For example, Fleming Peterson is now a senior at Wake Forest, John Macaroco transferred to the University of Rhode Island, and Stephen McMenaman is now at Stonehill College.

We don't have a recruiting budget or dorms here at Massasoit, but what we do have is a college any community would be proud of. I like to think I've helped.

Nick Palantzas graduated from Massasoit Community College in 1978 with a degree in physical education and now lives in Brockton, MA. His current occupation is assistant athletic director and mens soccer coach at Massasoit Community College.

Marissa Quiles
Waubonsee Community College, IL

Ican remember coming home from grade school with a stack of books piled so high that they blocked my view. There were evenings that I couldn't finish my homework, no matter how hard I tried, and my mother would be on the verge of tears because she couldn't help me. She didn't know English very well and was forced by her parents to leave school at an early age.

Throughout my early life, my mother's main concern was that I obtain a quality education and that I not suffer. Well, she had little to worry about when I decided to attend Waubonsee Community College.

I can truly say that one of the secrets of my success is Waubonsee Community College—with its buildings surrounded by trees and full of caring people. At first I was hesitant to attend Waubonsee Community College because I wanted to go away to a larger university. Lack of funds kept me at home, but going to Waubonsee turned out to be one of the best choices I ever made. It prepared me well for my career which began at the age of 21.

I am 24 years old now, successful and happy. I'm the only woman in top management at WSNS-TV Channel 44 in Chicago, as well as the youngest. I'm producing parades, beauty pageants, holiday specials, and more. These are goals that I never thought I would reach so quickly, and sometimes I fear they will end just as swiftly. So no matter what happens, I know that I have a strong academic foundation thanks to Waubonsee Community College.

Whenever I am asked to share my secrets of success to students, I always recommend that they attend a community college. I believe that a community college can lead you in the right direction, is affordable, competitive, and full of rich opportunities for hands-on experience. Most importantly, a community college makes you feel like you belong; it feels like home.

Many of my friends are successful accountants, reporters, policemen, artists, instructors, politicians, and with them I share this in common: we all attended a community college.

I want to thank Waubonsee for giving me the confidence I needed to succeed, for making the transition into a university and the real world easier, for teaching me about life, for telling me that I was special, for believing in me, for giving me the opportunity to grow, and for making sure that instructors Shirley Borel and Jan Sprague-Williams were there with me every step of the way.

I am happy to say that attending Waubonsee has become a tradition in my family, in which I am the oldest of five. I am glad that I set an example that my brothers and sisters are following.

Marissa Quiles graduated from Waubonsee Community College in 1986 with a degree in journalism and now lives in Chicago, IL. She works for WSNS-TV Channel 44 in Chicago as program director and executive producer. Her tenacity and drive, coupled with her outstanding career successes, have provided an excellent role model for everyone—especially women and minorities. She was recently awarded the "We Care" Role Model Award by the city of Chicago for outstanding role models.

Ruth Y. Radin
Hartford College For Women, CT

Beyond the intrinsic value of gaining knowledge, and hopefully growing in wisdom, I now see education as increasing one's opportunity for choice in life. But when I enrolled at Hartford College for Women to study liberal arts after my senior year in high school, I didn't have any notion I was doing this. I did feel somewhat disappointed that I wasn't going away to college as my friends did, but financially it was out of the question. In addition, HCW was located in a large old house at that time. The physical setting was hardly collegiate.

As my freshman year progressed, however, I realized that HCW was not like high school. I was expected to defend my positions, search for significance, make choices, and become meticulous in the way I expressed myself. I suspected that my English professor, Mr. Butterworth, put as much time into writing criticisms of my papers as I spent writing them. He was a writer himself, and that year his first book for children was published. Little did I realize that one day I would have my children's books published, too.

In addition to the impact my courses and professors had on me, it was soon apparent that if we, the students, wanted something to happen at HCW, we had to take some responsibility for making it happen. If we wanted to dance, we had to roll up the living room rug. If we wanted a yearbook, everyone had to work on it. We were encouraged to become independent and innovative. What we did really mattered.

With an associate degree from HCW, I transferred to Connecticut College and majored in music. After graduation, I became an elementary classroom teacher and earned my master's in education. While I was raising three children, I was the coordinator of their co-op nursery school and involved myself in other volunteer activities in the community, knowing that if I wanted something done, I had to help make it happen. As my family grew, I went back to college and became certified as a reading specialist. During that time, I became interested in writing for children and now have five children's books published. In addition, I am coordinating an adult literacy program in my community.

Education has given me the opportunity for choice in my life and the feeling that I can do much if I put forth the effort. This philosophy, as well as the academic means to accomplish it first took hold in the wonderful experience I had at Hartford College for Women.

Ruth Y. Radin graduated from Hartford College For Women in 1958 with a degree in liberal arts and now lives in Bethlehem, P.A. A writer of children's books, she has published A Winter's Place, Tac's Island, Tac's Turn, High in the Mountain, *and* Carver. *She has taught children in the third and fourth grades, coordinated a cooperative nursery school, and developed a course on writing children's literature which she taught at the University of Rochester.*

Ford Rainey
Centralia College, WA

Centralia College at its founding was the first then-called junior college in the state of Washington. Its principal founder was Margaret Corbet, a teacher of English in the high school. She cast me in my first roles in plays in the high school and influenced me not only to enroll at Centralia College, but to tour southwest Washington persuading other high school seniors to enroll also.

I continued acting in Centralia College under Margaret Corbet's direction. After attending a performance of George Arliss as Shylock in Shakespeare's "Merchant of Venice" in Seattle along with the cast of one of Miss Corbet's productions and realizing his magnificent performance meant so much more to me than to the other student actors, I decided then and there to become a professional actor. (George Arliss may be remembered by elders like myself as Disraeli in the motion picture of that name.)

At Margaret Corbet's suggestion, I auditioned and was accepted on a scholarship to the then famous Cornish School in Seattle. While attending Cornish School, I started working in radio along with Chet Huntley and Frances Farmer.

A professional repertory company was formed out of the Cornish School, and we toured the West Coast and finally crossed the country winding up in New York City.

My most prestigious part on Broadway was as "J.B.," the title role in Archibald McLeish's Pulitzer Prize play.

My first feature film role was as a member of Jimmy Cagney's gang in "White Heat." Many features followed and countless television shows—particularly westerns. I was the Six Million Dollar Man's stepfather and the Bionic Woman's guardian and am still active in features, television, and the stage.

My next feature picture may be released in 1990 for Academy consideration. It is "Bed and Breakfast," starring Roger Moore with Talia Shire and Colleen Dewhurst. I play Colleen Dewhurst's lover. (Comedy!)

Between engagements I still avail myself of the opportunity to attend community college classes, principally Santa Monica College, but also Los Angeles City College. I especially value the courses in piano, French, and voice at Santa Monica City College.

Without the influence of Centralia College, and particularly Margaret Corbet, I would never have dreamed of becoming an actor.

> *Ford Rainey graduated from Centralia College in 1930 with a degree in general studies and now lives in Malibu, CA. As a movie, television, and theater actor, his acting career has spanned over 55 years.*

Anne Marie Riccitelli
Community College of Rhode Island, RI

Before I was born, my father started to buy books for the children he expected. He was sure that his children would not begin work at the age of seven as he had done. We would have the priviledge of an education. He encouraged me to study, to learn to use language well, and to love books. While his friends discouraged him from educating his daughters (this was pre-women's lib, and women were supposed to marry and have children) he knew from his own experience that education was never wasted nor lost.

I wanted to go to college, but I didn't know how I'd pay for it. It was 1964, the year the Community College of Rhode Island (then called Rhode Island Junior College) was founded. For me, this was quite a stroke of good fortune. I was a member of the first class at CCRI. I was able to live at home for two years, save money, earn an AA degree, and transfer to a four-year college for a BA. Classes were small, teachers were dedicated and supportive of our dreams and ambitions. Everyone wanted to prove that CCRI students could successfully transfer to four-year schools and earn baccalaureate degrees.

I most remember being encouraged by my teachers to write and use language well. I studied liberal arts believing that the humanities were the basis for a sound education and the enjoyment of life. I worked hard, edited the school's fledging literature magazine, made the dean's list, and at the end of two years transferred to the New School for Social Research in New York. I later earned an MA from New York University.

I had no idea that my love of language would lead to my first job as an editorial assistant at *Harper's Bazaar*, and later to writing articles for numerous national magazines. Today, I am manager of magazine publicity for ABC Television. In this position I must communicate with important editors at major national magazines about ABC-TV's projects, programs, executives, and talent. I must present my company's point of view to tough, discerning people who are quick to criticize and comment—in print.

I'm grateful to my father who believed in his daughters as well as his son (all three of us started at community colleges), and to CCRI, which helped me realize my dream of a college education.

> *Anne Marie Riccitelli graduated from the Community College of Rhode Island in 1966 with a degree in liberal arts and now lives in New York City. She is manager of magazine publicity for the ABC Television Network Group. She is a member of the American Society of Journalists and Authors and of the International Women's Writing Guild. She also serves as a foundation trustee for the Community College of Rhode Island.*

Jay Rodriguez
Mt. San Antonio College, CA

At Mt. San Antonio College, I gained a fascinated interest in journalism, after enrolling in a class taught by Hilmer Lodge, the journalism instructor. Hilmer was a very special person. A track coach at the school, he was a very inspirational coach and teacher. He made the field of journalism so interesting that I changed my major, and then Hilmer also changed my life forever.

One day he announced that there was a job opening at the Pomona *Progress Bulletin* in the circulation department. He asked for a volunteer for the opening, and no one raised their hand.

After making assignments to the class, they started to work at their desks and Hilmer came to my side and asked why didn't I go to see the business manager of the paper, and inquire about the job.

So I went and accepted the job.

The job in circulation turned into a job in the advertising department. I started running proofs, and then advanced as a junior salesman and then a salesman of display advertising. That culminated in a job 10 years later as director of special editions for the newspaper. But I continued all those years to take courses at Mt. San Antonio College to learn new skills and about new fields.

After 13 years at the paper, I left to join two friends in the restaurant business. I did the marketing and public relations for five restaurants that we owned. I shifted the courses at Mt. San Antonio College to business courses, and continued to attend classes at night.

The restaurants had big-name entertainment and employed people like Louis Armstrong, Ray Charles, Ella Fitzgerald, James Brown, The Righteous Brothers, and Duke Ellington.

I went on to a job at KNBC, the local NBC station in Los Angeles, as administrator of community relations. Then I advanced through the years to manager of community relations and manager of press and community relations.

In 1979 I assumed the position of vice president of corporate information for NBC. My responsibilities include corporate advertising, corporate donations, corporate events, serving as the NBC representative with special interest, government, and industry groups, and coordinating the speakers' bureau and corporate projects.

The studies I took at Mt. San Antonio College prepared me extremely well. For example, I did not know then why I took some speaking courses at Mt. SAC, but in this job I have spoken to hundreds of groups around the nation and testified before a United States Senate Sub-Committee.

But the turning-point was that day in a college journalism class when a teacher took a special interest in a student and encouraged him to apply for a job he did not want.

Jay Rodriguez graduated from Mt. San Antonio College in 1956 with a degree in liberal arts. He is currently serving as vice president of corporate information for NBC and resides in Glendale, CA. A lifelong supporter of community service, he has continued to hold a special place in his heart for Mt. SAC. A recipient of the college's Alumnus of the Year Award, he represents a person who has achieved and remembers those who contributed to his success.

Jeffrey L. Scott
Kankakee Community College, IL

Upon graduation from high school in 1972, I had the opportunity to fulfill my lifelong dream of becoming a major league baseball player. I was drafted and signed by the Texas Rangers that June and left home for a summer in the minor leagues.

One significant part of my signing bonus was a provision for payment of my college education. Education was important to me, as well as to my parents, so during the summer of 1972 I put considerable thought into choosing a college at which to continue my education. I determined that, because of its proximity to my home, Kankakee Community College was the proper place for me to attend.

Due to the nature of the profession in which I was engaged, I had to attend college in a rather unorthodox manner. I could go to school only a semester at a time, each year when the baseball season was completed. From the fall of 1972 through the spring of 1976, I attended KCC in the winter months and then left each spring to go to spring training.

Because of the unique opportunities KCC offered at that time, I was able to attend a four-week winter term each January. Doing this allowed me to earn enough credits to graduate with my AA degree on schedule. I then transferred to a four-year college ready to tackle the rest of the education I needed to receive my BS degree.

I was also quite fortunate that at a community college I was able to get the attention and guidance I needed in light of the rather hectic approach I took to attending college. In fact, I had the opportunity of a lifetime. I was able to pursue my chosen profession and receive an education at the same time. Because of my time at KCC, I had the best of both worlds and was able to achieve many of my most important goals.

Jeffrey L. Scott graduated from Kankakee Community College in 1976 with an AA degree. He presently works as a professional baseball recruiter, helping young men deal with the pressures and demands they encounter as professional athletes. He has been a member of the Association of Professional Ball Players of America for 19 years, actively providing a support network for former athletes. He also is a member of the Pitch and Hit Club of Chicago, a nonprofit organization which supports youth baseball in Chicago.

Warren Skaaren
Rochester Community College, MN

College sneaked up on me like a Thanksgiving fog, the way it did for a lot of rural students. Focused three feet in front of my nose, dazzled by the urban social whirl of the high school senior, I worked in the evenings and studied hard. Cash was short, and although my grades were good, my foresight had yet to appear. So when it came time for college, I found myself alone, surprised, and without a plan. Thank God for Rochester Community College.

Frightened, I crept up to the arched doorway and was swept inside by enthusiastic sophomore counselors and a determined associate dean of students, Robert O. Wise, Jr. He schemed and cajoled until I attended a freshman camp in the forests of Wisconsin and there won the first thing I'd ever won in my life: a leg wrestling contest. Buoyed on the tide of such an Olympian victory, I sniffed the wind for headier battles. Dean Wise called me into his office and peered at me through clouds of ah ...inexpensive... pipe smoke and said "I'm going to give you the secret of life, son: PROJECTS!"

Certain that he meant building dams and bridges, I said I wanted nothing to do with construction work. But it turned out that Dean Wise meant that I should organize my life around projects that capture my curiosity. Possessed by a project, you can whet your appetite, set goals, attract other like-minded souls, and—win or lose—you'll have learned mighty lessons along the way.

Inspired, I ran for offices, joined the drama club, dated the Homecoming queen and was elected student senate president. Along with other caring faculty members, Dean Wise helped me confront a lack of personal confidence and a self-image that predicted, "You'll be a farm boy working at the canning plant until they wheel you away. Get out of the way and let city folks run the ship."

I matriculated and went, armed with new confidence, to Rice University, graduating with honors, as student body president there, too. I served in the Texas Governor's Office, and now I'm in the entertainment business working as a freelance writer, moving, you guessed it, from project to project. Although leg-wrestling turned out to be a limited skill, I have learned some mighty lessons along the way.

In the small, safe environment at RCC I found some inspiring, accessible teachers, the company of excellent students, and a dean to whom I still turn from time to time when I need a nice word.

Rochester Community College was the right place at the right time for a farm kid without a plan. I am deeply grateful.

Warren Skaaren graduated from Rochester Community College in 1967 with a liberal arts degree and now lives in Austin, TX. His current occupation is freelance writer. He is the writer of "Beverly Hills Cop II," "Beetlejuice," "Beetlejuice II," "Days of Thunder," "Fire with Fire," and "Crimson Eagle," as well as a co-writer of "Batman" and writer/director of "Fireworks," "A Special Place," and "Breakaway."

John C. Walters
University of Cincinnati-Clermont College, OH

The challenges facing a returning student are almost commonplace today. But when I started college, I wasn't just returning to school, I was returning to civilian life, returning to the country, and coming straight from a stint with the special operations in Vietnam. If you want an example of an at-risk student, look no more. But I not only completed the associate degree in business, I also continued with my education, became an instructor at the university, and now serve on the faculty of the Winona International School of Professional Photography in Chicago and publish a regular column in *Professional Photographer*, which is the official trade journal of the industry.

University of Cincinnati-Clermont College—and the special kind of institution it was and is—played a critical role in my ability to make that tough transition. Faculty had time to talk with students. A faculty member who started at Clermont about the time I did was a retired military officer; we shared war stories and talked about the experiences we'd been through. Finding time, treating adults like adults—that was the rule, not the exception. And the academic support was more than I could have ever expected. I had a tough time with accounting, and I can only describe the faculty as unwilling to let me fail. If you told an instructor you had a problem with something in the course or the college, it wasn't just your problem anymore; it was something the two of you were determined to overcome. With the strong background provided at Clermont, I had the best preparation for transferring to the main campus and a four-year degree.

I've had my own studio since 1974, have taught at the university for 10 years, and regularly speak, publish, and often win competitions at the national level. And I set standards for myself to be the same kind of concerned instructor as those I had years ago.

John C. Walters graduated from the University of Cincinnati-Clermont College in 1974 with a degree in photography and now lives in Amelia, OH. He is a professional photographer with membership in the Professional Photographers of America, for whose trade journals he has written more than 15 articles.

Dennis C. Webster
Jamestown Community College, NY

Three weeks into my freshman year of college, I faced the first crisis of my adult life.

I was lost and alone in the vastness of a major eastern university. While I was academically eligible, I soon discovered I was in no way ready to handle the social and personal pressures of a large school. I was one small person in a sea of students just like me.

In late September, frustration turned to agony. My father took ill, was hospitalized, and died within a matter of 72 hours.

I had thought before his death that the big school was not going to work for me. Afterward I knew. Confused and grieving, I struggled for an answer.

Jamestown Community College was the answer.

I completed that first semester at the university, barely passing, and transferred back home, to JCC, in January of the following year.

At JCC I found courses that would allow me to live at home and still prepare for an academic career. And JCC also gave me something else, something even more important. There were people who knew me and cared about me—English professors, philosophy professors, psychology professors who were accessible, who had the time to listen, and could help me sort out the feelings of failure and grief that overwhelmed me.

I studied and learned many of the things that the big school offered. At the same time I gained confidence and better self-esteem; I was happier.

At the conclusion of my sophomore year I graduated from JCC and went back to the major university setting where I eventually received my BS degree. Subsequently, I enrolled at another school in the State University of New York System, where I earned a master's degree in English linguistics. I now am the operations manager of an FM radio station.

As I reflect on that time at JCC almost 20 years ago, it is not so much the courses I took as what happened to me when I took them. I began to get a grip on my own life.

Dennis C. Webster graduated from Jamestown Community College in 1974 with a degree in English and now lives in Jamestown, NY. As operations manager for WJTN/WWSE Radio, he has become "the voice of Jamestown." His community knows it can trust Mr. Webster for accurate and up-to-date information.

JoAnn Young
County College of Morris, NJ

Remember the movie, "The Way We Were," when Robert Redford and Rip Torn name their "best years"? Not counting the obvious marriage and birthdates, I can name two of my own.

There was August 5, 1983, when I got the call that I was nominated for an Emmy for writing and producing a television documentary about Mario Lanza. Family and career—that day I had it all!

The other day, February 4, 1974, is directly responsible for that memorable August afternoon. I had enrolled at County College of Morris as a full-time student at age 35, which seems young now. At the time, I felt as ancient as the 1956 high school graduate I was. My three classes that day were world history, American literature, and drama. At each class the professors outlined study for the semester. It was like being shown a delicious buffet of your favorite foods, and the assignment was to eat! By the end of the day, I was lightheaded with excitement.

The next day there were more scholastic adventures and, honestly, my enthusiasm stayed very high, even with the pressure of being a wife, mother, and aggressive student. CCM accepted all my (much) earlier credit hours from the University of Nebraska, so after two semesters and a summer session, I had earned an AA degree. What's more, CCM gave me the encouragement, confidence, and direction to get a BA in 1977.

Deciding to resuscitate my childhood dream of working in the broadcasting field, I started the grind of resumes, networking, and interviewing. One Sunday after watching a CBS program called "Camera Three,"—a show primarily about the arts, sciences, entertainment— I wrote the executive producer telling him what I liked about the program. The day my letter arrived, the production secretary transferred to another show, and the producer said, "Call this woman." That was the beginning of my second career.

Today, I am a writer-producer of documentaries and entertainment programs for public television, working with major talents of yesterday and today, from Liza Minnelli to Lawrence Welk to Andrew Lloyd Webber. We did seven programs at the Reagan White House and two for President and Mrs. Bush. Next March, "Music by Richard Rodgers" will be broadcast nationally on PBS, a contrast from this year's "John Wayne Standing Tall."

A third important day in my life occurred when I drove to CCM and asked a very nice woman about taking some classes. She was the one who told me, "You can do it" and encouraged me to sign up as a full-time student. That wonderful woman and CCM absolutely changed my life and opened the door to the career that I dreamed about as a star-struck kid in Nebraska.

> *JoAnn Young graduated from County College of Morris in 1975 with a humanities degree and now lives in Oradell, NJ. She is writer-producer of public television documentaries. Her documentaries and entertainment specials feature major talents of yesterday and today.*

CHAPTER SIX

Health and Medicine

Roger Dale Burress, Sr.
Roane State Community College, TN

I grew up in a small, rural Tennessee community. Neither of my parents have a college education, and at the time few of us felt any particular push to pursue a college degree. I had done fairly respectable work in high school academically, and I found myself in the top 10 students in my graduation class. When I interviewed with the private universities that I wanted to attend, the universities saw deficiencies, specifically deficiencies in foreign language, which our school didn't even offer at the time. But the community college I investigated, Roane State Community College in Harriman, TN, was willing to accept me as I was and provide me the opportunity to build from this foundation. In this community college, however, I felt the competition with students who were trained at larger high schools where the facilities and opportunities were greater. I had to learn how to set goals and keep my eyes firmly fixed and overcome the obstacles to those goals. Usually, I have found that overcoming my obstacles involves changing something about me or changing my approach to the task rather than my surroundings. Later, when I did transfer to a prestigious private university, I was able to take a degree in a challenging major and go on to complete my medical training. I was president of Gamma Beta Phi, the honor society at RSCC, and was involved in drama at the college. The faculty and administration at that institution were probably more responsible for my ultimate further endeavors than any other institution. I met a lot of good people there and received a lot of good encouragement from the faculty and staff.

Roger Dale Burress, Sr. graduated from Roane State Community College in 1979 with a degree in pre-medicine and now lives in Oneida, TN. He has become an internal medicine specialist in Oneida because it suffers from a shortage of doctors. He is a licensed auctioneer, founder of the Big South Fork Auction Company, and a licensed real estate affiliate broker with Tri-Mountain Realty Company in Wartburg, TN.

Michael Cochran
Southwest Virginia Community College, VA

During my childhood, my parents stressed the importance of a good education. With this in mind, I always enjoyed school and being with other people. Education was important in my life, allowing me the opportunity to go places, do things, and meet new and important people through the miracle of study and learning.

I was excited when I received news that I was accepted in the radiology technology program at Southwest Virginia Community College. Being accepted to this program, I could continue study in a science curriculum, my favorite subject area, as well as build a career where I could work with and serve people.

While studying and learning during the 27-month program, I enjoyed the challenging classes ranging from English and speech to psychology and government. The radiation sciences introduced me to the world of x-ray production and protection while allowing further in-depth study of physics, anatomy, and math. However, as I was to learn upon graduation, all these classes molded my background into a career of serving and aiding people.

After graduating with the skills I had learned and achieved, I began work as a registered radiologic technologist on staff in a radiology department. It was challenging to know that by producing desired radiographs, I was helping to diagnose injury and disease. With the desire to continue to learn, I continued to progress in my career. I next moved to a job as a special procedures technologist, progressing to a computerized axial tomography technologist, next to magnetic resonance imaging supervisor, then to cardiac catheterization technologist, knowing that with each job move, I was utilizing what I had learned to move deeper into advanced technology.

Progressing technically, I needed to become involved in professional societies that stimulate growth as well. Having joined the American Society of Radiologic Technologists in school, and being recognized by the American Registry of Radiologic Technologists upon graduation, I joined my local and state societies as well. It wasn't long before I became active in both. In the local society, the SWDSRT, I assisted for two years with the business aspects of the organization, then ultimately was elected its president. In the Virginia state society, I served on several committees and assisted with a convention, as our district was host to the annual meeting.

Still, after being in the profession of radiologic technology for 14 years, each day continues to be a learning experience, just as enjoyable as the first. The thought of being able to help a fellow human, and the possibility of saving a life, is the greatest reward anyone could ask for. Never wanting to say I have learned all I can, I continue to look at each new day as a means of continuing my education and broadening my horizons.

Michael Cochran graduated from Southwest Virginia Community College in 1977 with a degree in radiologic technology and now lives in Richlands, VA. As a radiographer, he has been extremely active in community and professional organizations.

Catherine Collins
Trocaire College, NY

My parents are loving and supportive, and wanted all of their children to have a college education. However, like most inner-city residents of the 1960s, they could not afford it. Of course, this was disappointing, and another route to complete my education had to be sought. I chose to enroll in the federal government manpower training program for licensed practical nursing. Once I had completed this course, the thirst for a college education continued to haunt me. I had heard about Trocaire, a small two-year college that provided an excellent education for registered nurses. I decided to apply, and Trocaire lived up to its reputation. The years I spent at this college were the most rewarding of my life.

At Trocaire College, the staff and faculty helped me to adjust to college life and academics. This junior college provided an excellent academic foundation upon which I have successfully completed a bachelor's degree in vocational education and a master's degree in allied health education. I am currently completing a doctorate in education administration at SUNY Buffalo.

Since graduating from Trocaire College, I have held primarily administrative positions. Currently, I am full-time assistant academic dean at Erie Community College and a part-time assistant professor at SUNY Buffalo. I have been cited by New York Gov. Mario Cuomo and the Secretary of the United States Deptartment of Health and Human Services for the Attica Prison Health Program. In 1989-90 I received one of SUNY's Best Faculty Fellowships, which allowed me the opportunity to expose New York City minority youth to the benefits of a junior college education. Thank you, Trocaire College.

Catherine Collins graduated from Trocaire College in 1970 with a degree in nursing and now lives in Buffalo, NY. She is currently full-time assistant academic dean at Erie Community College and part-time assistant professor at SUNY Buffalo. She has been cited by New York's Governor Cuomo for her involvement in the Attica Prison Health Program.

Carlos Estolano
Southwestern College, CA

I graduated from a high school in Tijuana, Mexico, at age 16. I wanted to learn how to speak English, and my best bet was to enroll in Southwestern College in Chula Vista, CA. My father lived in the area, and I moved in with him. My mother, who lives in Mexico, motivated me to continue in school. I worked part-time in my father's business (sewing leather jackets) while I attended college.

I initially took English, chemistry, and math classes. I never realized how challenging my schedule was going to be. I rapidly noticed how many of my Southwestern instructors invested their time attempting to help me learn, despite the language barriers.

The Southwestern College career advisers helped me evaluate my decision to apply for the nursing program. My control of English was not that great when I first entered the nursing program. However, the instructors were extremely helpful during the two years of training. I was also very fortunate to find my best friend (in the same nursing program), as he was my tutor, and at times my translator as well.

I passed the State Board nursing exam on the first attempt—a result of my nursing instructors' persistent interest in my success.

My first job as an RN was to become the night-shift supervisor of a 192-bed convalescent hospital. I was the only RN in the facility during this time. The responsibilities were enormous. I now understand why my instructors at Southwestern College were so intent on trying to teach us as much as they did during our training.

I joined the County of San Diego Mental Health Services Division in April 1984. I was a medication nurse initially and am now the head nurse of the Emergency Psychiatric Unit for San Diego County.

My duties include the 24-hour planning and coordination of safe patient care, administration of nursing services, and proper staffing of the unit. I supervise other professional nurses, and I serve as the link between the hospital administration and the nursing staff under my supervision.

While I was a student at Southwestern, I did a training rotation in this hospital. I never imagined that one day I would be managing one of that same hospital's units.

The knowledge I have gained during my work in the County Psychiatric Unit has trained me for crisis situations. I recently assisted victims of the San Francisco Bay Area earthquake disaster.

The foundation of knowledge I gained while at Southwestern gave me the needed tools to become successful in life. Moreover, I have found nursing a very satisfying profession.

> *Carlos Estolano graduated from Southwestern College in 1983 with a degree in nursing and now lives in San Diego, CA. He is head staff nurse of the emergency psychiatric unit at the San Diego County Psychiatric Hospital.*

Lucia F. Ferguson
Greater Hartford Community College, CT

As a divorced and single parent, I often wondered what the future held for me and my son. The year was 1976, and I had been out of high school for eight years. For the last two of these, I was a welfare recipient and had become well acquainted with the social service system. Could I possibly be of help to others who shared my situation?

Greater Hartford Community College was offering a certificate program in social services, and I enrolled, not certain where it would lead. I did so well that I was encouraged to enter the liberal arts associate degree program, with a major in sociology. My professors made me want to reach for more. Finally, I was being credited with having an original thought!

During my community college years, I was fortunate enough to become part of a program providing hands-on training in substance abuse counseling. Over the past 12 years, I have used my skills in substance abuse prevention, intervention, and treatment as a counselor in an out-patient drug-free facility, a methadone clinic, an in-patient detox hospital, and a residential hospital. Currently I am an AIDS/HIV program coordinator for the Connecticut Alcohol and Drug Abuse Commission.

In the decade since I became a proud community college grad, I've worked hard to complete my bachelor's degree. I am now enrolled at the state university, studying toward a master's degree in public health. I also earned a state certificate as a substance abuse counselor.

While I continue to move ahead, I always credit my first experience with the community college system for motivating and encouraging me to want more for myself and my family. Today I also am involved as a board member in the Alumni Association of Greater Hartford Community College. This I do with pride. It is important for me to give something back to the college that gave me and countless others so much.

Whenever I feel like a number in the state university system, I remember my community college roots and say, "Thank you, Greater Hartford Community College, for giving me the courage to believe in myself."

Lucia F. Ferguson graduated from Greater Hartford Community College in 1978 with a degree in liberal arts. Her current occupation is AIDS/HIV coordinator in the Connecticut Alcohol and Drug Abuse Commission in Rockville, CT.

Barbara J. Gade
Lexington Community College, KY

I will never forget the thrill of being accepted into the first dental hygiene program at Lexington Technical Institute (now Lexington Community College). I had been preparing for admission by taking basic coursework at the University of Kentucky. I was somewhat hesitant, since I was 38 years old, and becoming a full-time student again was foreboding.

The first year was difficult in many ways; I had some doubts about surviving. However, once I learned to study again and to still maintain a life outside of the classroom, things soon fell into place. To my pleasant surprise, the hygiene class evolved into a homogenous and congenial group. Also, at LTI I found a competitive side of myself that I didn't realize existed. My professors nurtured and challenged this trait; it was exciting to be learning and achieving my long-time goal.

After graduation I took a position as an Instructional Specialist II, teaching expanded duties to dental auxiliaries. This experience "hooked" me to return once again to evening classes at the University of Kentucky to seek my bachelor's of health science degree. I found that teaching is another way of learning. It is a thrill to see another person succeed by learning new skills and concepts as I guide them along the way.

In my present position as a dental hygienist at the Veterans Administration Medical Center, I have the best of two worlds. I practice my dental hygiene clinical skills and am still able to work with dental hygiene students who rotate through our Dental Service, thereby giving me the opportunity to maintain my teaching skills and interests.

The confidence to succeed was nurtured by my husband and the professors at LTI. I am grateful for the encouragement and support that I have always received. It has added immeasurably to my personal growth. I would like to express my gratitude to each and every one for all their help in reaching that goal.

> *Barbara J. Gade graduated from Lexington Community College in 1979 with a degree in dental hygiene and now lives in Lexington, KY. Her current occupation is dental hygienist at the Veterans Administration Medical Center in Lexington. She has received numerous awards, including the "Outstanding Federal Employee" for 1988 at the Veterans Administration Medical Center.*

Reyna Garcia
Community College of Denver, CO

Community colleges are often overlooked or not given the credit they deserve. My experience at the Community College of Denver was one of the most positive points in my life.

I entered CCD at a very unstable stage in my life. I had dropped out of school when I was 14 years old. Soon after, at the age of 15, I became pregnant. My daughter was 10 days old before my 16th birthday. If this sounds like a grim story, it's because it was.

Not wanting to fall into the welfare trap or depend on my parents to help support my daughter, Gabrielle, and me, I decided to get ahead by way of education. Earning my GED was the first step, but I soon realized that I would need more than a high school equivalent certificate. I decided to take the next step—entering college. I decided to attend a two-year college because my need was more immediate than a four-year college could satisfy at this point.

With help from grants, scholarships, and student loans, I was on my way. I remember walking into CCD for the first time, though, and it was a bit overwhelming for me. My self-confidence was low, but because of caring counselors and teachers, I soon felt at home in my new environment.

I was not sure exactly what I wanted to major in, but support and suggestions from teachers and counselors enabled me to choose a field I liked. They laid out my classes for the medical secretary program; it was a tight schedule, but with my new-found confidence, I was sure I could do it. My last two semesters I had an overload schedule (22 credit hours), tutored students in two classes, worked as an assistant in the secretarial lab, attended Phi Theta Kappa meetings (a fraternity for honor students), worked as a part-time volunteer at St. Anthony's Hospital, and cared for my daughter after school. Towards the end of the program, I was exhausted, but I knew my goal was now within reach.

Graduation finally came. Standing on the podium accepting my degree and two certificates, I had feelings of complete satisfaction. I was actually very proud of myself; I felt as though I could meet the challenges in life that lay ahead.

Soon after I graduated, I landed a job at St. Anthony's Hospital in the Medical Records Department. It was an entry-level position, but it meant so much more to me. I have been there for one year now, have had a promotion, a great raise in salary, and I'm completely self-sufficient. I moved out of my parents' house with my first paycheck and am now able to afford rent, daycare, and other necessities of life.

Reyna Garcia graduated from the Community College of Denver in 1987 with a medical secretary degree and now lives in Denver. A medical secretary, she serves as a member of the Secretarial Program Advisory Committee and recently volunteered to work with students at West High School on behalf of CCD programs.

Many great things happened because I once found the courage to walk into CCD. Not only did I gain a good education, but I acquired self-confidence, motivation, and the respect of wonderful teachers along the way.

Brenda Sue Gierhart
Brookdale Community College, NJ

I t was a good match. I couldn't have chosen a better college for what I needed at that time in my life. It was close and inexpensive—no small matter when you are 17 years old and still living at home. My choice was to attend Brookdale—or not to attend college at all. I am now a busy obstetrician-gynecologist caring for many patients with complicated medical problems. Yet it might have been quite different. When I go to the hospital and work with a nurse's aide, I think, that was me "half a lifetime" ago. And I might still be a nurse's aide—except for Brookdale.

I was eager to learn, and Brookdale was ready to provide me with a sound education. I met professors who were excellent teachers. In fact, I rank my Brookdale anatomy professor, Frank Gimble, far ahead of my anatomy professors in medical school. Many of my teachers took a personal interest in me, encouraging me to continue my studies and serving as mentors to help guide the path my life would take.

The clinical experience that I gained as a nursing student at Brookdale put me years ahead of my fellow medical students. And working with the patients motivated me to learn even more—so I could give them better and better care. There were many dark moments along the way. Yet Brookdale had given me a strong educational foundation to build on, and it had given me a series of successes that gave me the self-confidence I needed to carry me through my BSN at Rutgers-Newark, my pre-med courses, and into medical school.

Brookdale also gave me the means to financially support myself through those years. I was able to pass my LPN boards after completing the first year of nursing at Brookdale. That raised my earning potential, so I could work fewer hours and still make the money I needed. And upon graduation from Brookdale, my salary jumped again as I worked as an RN. Even in medical school, Brookdale continued to help me with scholarships from the Brookdale Foundation.

When I was in London studying cardiology, I explored the city proudly wearing my T-shirts from the Brookdale Bookshop. The British people seemed fascinated with initials and would always ask me what the "BCC" stood for. So I would tell them and explain that "you can go a long way coming from Brookdale Community College."

Brenda Sue Gierhart graduated from Brookdale Community College in 1973 with a degree in nursing. She is now an obstetrician/gynecologist and resides in Indianapolis, IN. She decided to pursue a career in a vital field which many others, unfortunately, are avoiding for fear of legal complications. She has received a Medical Perspectives Grant from the National Fund for Medical Education to study hospice care for dying cancer patients.

Carmen P. Gioia

Community College of Allegheny County-South Campus, PA

It was always my dream to attend a major college. Often in high school, when conversation would turn to life after graduation, I would tell all of my friends about going to Princeton, Purdue, or Duquesne. But I knew that was only a dream. Although our parents always found the resources to provide what we needed, I seldom knew the pleasure of any "extras."

In 1973, realizing there would be no major college in my future, I began a two-year journey toward an associate degree in biology. Along the way my work as a lab technician and a student senator demonstrated hidden abilities that might otherwise have been overlooked.

Had I been in my "dream school," I doubt seriously my instructors would have been as human and caring as at Community College of Allegheny County. The "people-are-important" attitude of all of my instructors gave me the incentive to push on.

It was the "you-can-do-it" assurance from CCAC that helped me leave behind my insecurities and feelings of inferiority. Without that constant attention, I could not have gone on to Gannon University, graduating in the top 10 percent of my class, and on further to Palmer College of Chiropractic.

Today, I am known as Dr. Gioia. Several thousand people count on me to ease the pains that crop up in their daily lives. Several of them have had life-saving experiences in my office, but each of them owes their good health to my skills as a doctor, the fine institutions of Gannon and Palmer, and especially to CCAC, the force that drove me to success.

Carmen P. Gioia graduated from the Community College of Allegheny County, South Campus in 1975 with an associate's degree in biology. He is now a chiropractic physician and resides in Jefferson Borough, PA. As a practicing physician and educator teaching anatomy and physiology at CCAC since 1985, he has had the opportunity not only to heal the bodies of his patients, but also to offer advice and counseling to all those who need someone to listen. Outside the office, he has been regularly invited by local social and civic organizations to address their meetings on the power of positive thinking. His research and initiative have resulted in several patents being granted for ideas and inventions that have furthered the field of chiropractic.

Ann Grant
Cuesta College, CA

When we moved to rural San Luis Obispo from Manhattan in the late 1960s, I found myself unable to work in the occupation for which I had been prepared. There was a glut of high school English teachers in California, and no opportunity, now that our two daughters were school age themselves, of resuming work in that area. After a year of futility investigating other possibilities, I learned that there was a highly regarded registered nursing program at Cuesta College. Nursing and medicine had always been an interest of mine, and I decided to enroll.

I had some initial concerns about how thoroughly a community college could prepare students for a profession as demanding as nursing. I also wondered how the level of instruction would compare with that which I had received at the university level. I very shortly discovered that I had to work very hard to achieve the level of performance demanded in both the general education and nursing classes. It was rigorous preparation indeed.

After graduation, I worked in many hospital departments, including obstetrics, pediatrics, and medical and surgical nursing. One summer I had the opportunity to teach students operating room nursing, and I found that I could combine my nursing and educational preparation in a very satisfying way as a nursing instructor.

Since that time I have completed bachelor's, master's, and most recently a doctorate in nursing, and am now the chairman of the Division of Nursing and Allied Health at Cuesta College. This spring, Cuesta will be collaborating with a neighboring school, Allan Hancock College, to open a new LVN-RN completion program at that campus, and we are looking forward to that joint undertaking.

None of this would have been possible without the opportunities provided by my local community college. Cuesta allowed me the chance to make a mid-life career change that has substantially altered the rest of my life. It has given me the opportunity to make a contribution to my community which I would not have envisioned 15 years earlier. As I see nursing students graduate each June, I can tell them that their community college education provides a basis for tremendous opportunities, limited only by their own vision and determination.

Ann Grant graduated from Cuesta College in 1975 with a nursing degree and now lives in San Luis Obispo, CA. She is director of nursing and division chair of nursing and allied health at Cuesta. Having served as a member of the California Nursing Education Commission, she is now president-elect of the Southern California Associate's Degree Deans and Doctors and a member of the Special Commission on the Nursing Shortage in California. She has been published in the areas of AIDS education and ethics in nursing.

Becky Grimm
John Wood Community College, IL

In 1978, my senior year of high school, I was undecided as to what to do with my future. My counselors advised college, but since I didn't have a career chosen, they left the decision of which college to me. My parents and I agreed it would be wise for me to stay at home and test college life right here at John Wood Community College.

I was afraid I would be a little fish lost in a sea of students. To my surprise, my classes were small and very closely guided by the professors. You could always receive one-on-one instruction if needed. The camaraderie of the students was reassuring. The enthusiasm for knowledge and achievement was contagious; it made you hungry to attain goals and then set them higher. I was encouraged by my professors not to limit myself. They guided me through and gave me confidence to set my ambitions on being a doctor of chiropractic. I received my associate degree, transferred to Quincy College, and received a bachelor's degree in 1982. I was accepted at Palmer College of Chiropractic. At Palmer, the nurtured enthusiasm I received at John Wood Community College was enhanced, and I came to fully appreciate the guidance and advice I had received from my professors and fellow students. While at Palmer I joined the Sigma Phi Chi Sorority Alpha Grand Chapter and served as Ki-Kirist and Junior Kitrus for one year each, and I assisted the sorority's Grand National Council.

Since graduation from Palmer College of Chiropractic in 1986, I have started my private practice back in my hometown of Quincy, IL. I have served as president of the Great River Chiropractic Society, and I am a director of the Quincy Women of Today Chapter of a national community-service organization.

I am thankful to John Wood Community College for instilling a zest for knowledge and an enthusiasm for setting goals. My professors guided my years there and helped me develop and focus on a career for my future.

Becky Grimm received two degrees—an AA in 1980 and an AS in 1982—from John Wood Community College. She is now a chiropractic physician in Quincy, IL. In addition to her work, she is active in community affairs, especially through the Quincy Women of Today. This group coordinates a Block Mother program, provides an annual dinner for clients at the local mental health clinic, and "adopts" a needy family at Christmas. She also coordinates local artists' displays at her clinic, is an active volunteer at her church, and has directed operations of the Heart Association Gala Ball fund-raiser.

Sharon R. M. Heist
Greenfield Community College, MA

When I first climbed the steps to Greenfield Community College, it was with great anticipation and more than a little trepidation. After all, I was in my 30s, and college had seemed an unattainable goal for all of the years since high school. Could I do the work? Was I too old? Would I be embarrassed at being with all of the younger, better educated students who would be in my classes?

In the first hour of class, I lost my fears and gained a great joy that will stay with me the rest of my life. I will graduate from Smith College in May 1990 with honors, but that accomplishment is built on the foundation I developed in Greenfield. I have always loved to read, but at Greenfield I found inspiration in the teachers and other students whose joy was also in learning. Greenfield reinforced what I had already felt, that educational opportunity is the greatest gift anyone can give. It opens all of the other doors, and when the educational environment is as supportive and challenging as the one at Greenfield Community College, then we are only limited by our imagination.

After graduation, I will return to the Southwest, to the Navajo Reservation and to graduate school. Greenfield showed me my potential, gave me the courage to try new directions, supported and encouraged my efforts, and challenged my intellect. I have grown in so many ways, there is not room here to begin to describe them. My return to the Southwest has been a long-time goal and will soon see fruition, and it is in great part directly due to what I learned here. I am not alone; many of my friends are seeing personal achievement they never would have believed possible and all directly related to community college educations. The service they supply cannot be overstated. I have been fortunate to attain several honors and awards in the years since I began at Greenfield, and I truly feel that all of them should include the name of GCC in the honor, for here was my genesis. I cannot thank Greenfield enough for all it has given me.

> *Sharon Heist graduated from Greenfield Community College in 1987 with a degree in nursing and now lives in Charlemont, MA. At present, she is in her senior year at Smith College studying psychiatric nursing. She was an Ada Comstock scholar, transferring to Smith to continue her education. She was a Dana Fellow and has been cited for her work with Native Americans.*

Bonnie Hill
Rowan-Cabarrus Community College, NC

From the time I was a little girl, I dreamed of being a nurse. After my daughter entered school and my husband had his career underway, I felt it was time to make my dream a reality. I suppose I've always been the type of person who decides to do something and charges ahead full steam...and this time was certainly no exception. It was July, and the fall quarter at Rowan Cabarrus Community College was to begin in two months. When I called to request an application for the associate degree nursing program, I was told that there was already a waiting list for the present year and I would be placed on the following year's list, but was asked to come in for an interview. Well, I believe in the power of prayer and I knew if it was meant for me to enter school that year, a way would open. To this day, I am not sure what happened, but two weeks later I was called and informed that I had been accepted as a first-year nursing student.

During the next two years, I fell in love with nursing. Courses at RCCC were challenging, and my love for learning was stimulated. My professors helped me to realize potential I never knew I had, and in 1980 I graduated with high honors.

After graduation, I worked as a hospital staff nurse, a role for which I felt well-prepared with my ADN education. My love for nursing grew, and I became fascinated with the nurse practitioner concept. The "education itch" had started again, and over the next four years, driving over 50,000 miles, obtaining a BSN from UNC-Charlotte and an MSN from UNC-Chapel Hill, I became certified by the American Nurses Association and the North Carolina State Board of Medical Examiners as a family nurse practitioner. During the many long trips back and forth to school, I kept my eye on my goal and not on the miles getting to the goal, wisdom given to me by a mentor at RCCC, which I have since passed to my daughter, now a college student.

To say that I love what I am doing is a gross understatement. My work is challenging and rewarding, and I have tried to give back some of myself through volunteer community work which draws from both my nursing background and newly developed skills. I know that I am doing what I was meant to do, and I will forever be indebted to RCCC for challenging me to grow intellectually and personally. People ask me if I'm going to "go all the way" and get a medical degree. I reply, without hesitation, "I am all the way...I'm a nurse, and that's the best anyone can be."

Bonnie Hill graduated from Rowan-Cabarrus Community College in 1980 with a degree in nursing and now lives in Salisbury, NC. She is a family nurse practitioner.

Brenda White Hoggard
Roanoke-Chowan Community College, NC

I remember being scared and unsure of what I was about to do when I first walked through the front doors of Roanoke-Chowan Community College in 1985. I was 34 years old, a wife, and a mother of three. I also was unemployed, as the hospital where I had been working as a licensed practical nurse had recently closed. I knew that I had always wanted to be a registered nurse and that RCCC had one of the best nursing programs in North Carolina. I did not know, however, if I could cope with all the responsibilities of college studies, a home and family, my church and community activities, plus find another nursing job.

I arrived at RCCC just as registration had ended for the day. I thought that perhaps this might have been a indication that maybe I wasn't doing the right thing. However, a friend of mine who works at the college saw me and stopped to talk. I told her of my fears of returning to school after so many years out of the classroom. She assured me, though, that I was taking a step in the right direction, and she practically led me by the hand to the college personnel I needed to see. They too advised me of the proper channels to take and assured me they would help in any way possible. The caring attitudes and positive reinforcement I received from the college staff and faculty were exactly what I needed. That night I discussed with my family the pros and cons of returning to school. They could not have been more supportive—then or throughout my two years of study.

Indeed, my nursing studies at RCCC were not easy, but I knew when I graduated that I had received the best education possible. I know, too, that if it had not been for my family, instructors, friends, and the nearness of RCCC to my home, I would not be where I am today. I began working for Roanoke-Chowan Hospice, Inc. after graduation and had served as a volunteer one year prior to my graduation. In September 1989, I became director of the Hospice. I love it!

My heartfelt thanks go to Roanoke-Chowan Community College and its people for helping my love and knowledge of nursing grow to their greatest height.

Brenda White Hoggard graduated from Roanoke-Chowan Community College in 1987 with a degree in nursing and now lives in Windsor, NC. Her current occupation is registered nurse and director of Roanoke-Chowan Hospice, Inc. She is making a significant contribution to society by directing a staff of nursing assistants, licensed practical nurses, registered nurses, social workers, chaplains, and volunteers. While attending RCCC, she served as president of her nursing class and was a leader among her peers.

Patricia M. Hurley
County College of Morris, NJ

Iam pleased to applaud County College of Morris for its profound impact on my life and, by association, the lives of a thousand health care professionals.

In Fall 1968 I was a homemaker with three school-age children when CCM opened its doors and accepted me in the nursing program. After two years in one of the most innovative programs in the country, I graduated in 1970, and worked in medical/surgical nursing. I joined CCM as an instructional assistant in nursing and completed my baccalaureate degree in 1974. Wanting to continue teaching nursing, I received a National Institute of Mental Health Nursing Fellowship at New York University; then I completed my master's degree in nursing with a functional minor in curriculum development in 1976 and my Ph.D. degree in nursing research in 1978.

These four years completed my metamorphosis to a professional committed to molding students into nurses who cared about people.

Subsequent training led to a private practice in family therapy. Appointed to the NYU nursing faculty in 1979, I have had the opportunity to affect the personal and professional lives of students in the baccalaureate, master's and doctoral programs. I received promotion and tenure in 1985.

For the past four years I have played a major national role in the education of health care professionals on the Human Immunodeficiency Virus (HIV) epidemic. In 1986 I was the only nurse in the country who applied and was funded for three years by the National Institute of Mental Health to conduct HIV/AIDS educational programs for health care professionals of all disciplines in New York City. In 1987 I wrote a proposal to establish an AIDS Regional Education and Training Center at NYU to extend the HIV/AIDS education programs to all health care professionals in New York City, New Jersey, Long Island, Puerto Rico and the U.S. Virgin Islands. This proposal was funded by the Health Resources and Services Adminsitration for three years. Both grants generated almost $5 million in funding and educated over 10,000 health professionals in continuing and generic educational programs. As a result, I was invited to testify before the Presidential Commission on the HIV epidemic and the U.S. Senate Committee on Governmental Affairs for the hearing "AIDS: Health Care Services."

I often wonder if the CCM faculty members realize their impact on my life, the lives of

Patricia M. Hurley graduated from County College of Morris in 1970 with a degree in nursing and now lives in Ridgemont, NJ. She is an associate professor at New York University and was the first nurse in the county to apply for a grant to conduct HIV/AIDS educational programs for health care professionals. Her counsel and advice have been sought by the Presidential Commission on the HIV epidemic.

my family, people with AIDS, and health care professionals working desperately hard to care for them. The ripples in the pond do indeed spread far and wide. Thank you from all of us.

Willie Clenzo Johnson
Cosumnes River College, CA

I remember very well my Cosumnes River College experience. As I contemplated changing my major from biology and pre-med, I spoke with one of my chemistry professors. He took time to get to know me as a person. Instead of accepting my drop card like another number, he spoke to me for quite a few hours, pointing out my good points and those areas that needed improvement. He felt I had the potential to become a good doctor and that I should never lose sight of my dreams.

He was just one of the people at the college who took the time to guide me. My interest in the sciences led me to major in biology, but I learned more than science and math. I learned quickly that management of time was also very important.

Since my arrival at Cosumnes River College, I had felt somewhat uneasy about my future. But many of the faculty and staff were encouraging and quite supportive. I worked part-time as a chemistry laboratory technician at the college and worked full-time at a local canning company. I graduated in 1974 and transferred to the University of California at Davis, where I participated in the multi-ethnic program and majored in mammalian physiology—always keeping in mind those dreams I formulated at CRC.

Upon graduation from UCD in 1976, I volunteered at the UCD Medical Center and worked at Kaiser Permanente Hospital in Sacramento. While there, I learned the inner workings of a large metropolitan hospital and had contact with nurses, physicians, and patients. From there I did one year of postgraduate work at Creighton University in Omaha, NE, prior to being accepted to the University of Minnesota Medical School. While there, I participated in the Student National Medical Association. I also worked at the local chapter of the Sickle Cell Anemia Foundation in Minneapolis.

After graduation from the University of Minnesota in 1983, I did my family practice residency at Hennepin County Medical Center in Minneapolis from 1983-86. I am now home in Sacramento in a family practice and back at Kaiser Permanente.

Since returning, I have begun giving back to the community and to CRC what they have given me. I have participated as a speaker on medical issues in the community at schools and churches, and I am currently teaching as a preceptor for medical students at the University of California, Davis.

Also, I have been asked to be the commencement speaker for CRC's 1990 graduating class. I owe a great deal of my success to what I have learned from my community college experience at CRC.

Willie Clenzo Johnson graduated from Cosumnes River College in 1974 with a degree in pre-med and now lives in Elk Grove, CA. His current occupation is doctor of family practice. Even while attending school, Dr. Johnson was active in volunteer work with such organizations as the Sickle Cell Anemia Association. Now he is a respected physician and spreads the word on local and national medical and health issues in the community. He is currently serving as teaching preceptor for medical students at the University of California-Davis.

Stephen M. Kalafut
Mid Michigan Community College, MI

After graduating from Tawas Area High School in Tawas City, MI, in 1974, I decided to work for a few years before attempting college. In 1975 I began working for Tawas St. Joseph Hospital as an orderly and ambulance attendant, and this job started me thinking about a career in health care. After observing many jobs in the setting, I decided to pursue a career as a registered nurse because of its diversity in the health care fields.

After moving to Midland, MI, in 1977 and working for a local ambulance service, I had an opportunity to go to Mid Michigan Community College to obtain certification as a paramedic. As MMCC's first paramedic program, the coursework was more than I expected in the pre-hospital setting (1,000 hours of didactic and clinical work) and intrigued me even more about my future in health care.

Marriage and the birth of my first daughter, Rebecca, only slightly delayed the pursuit of my goal to be an RN. In 1984 I found myself at a point in life when I could balance family, job, college, and life in general. MMCC's associate degree in nursing program offered the flexibility I needed to pursue my dream. MMCC was less than one hour from my home, so classes could easily be taken, and clinical settings were likewise within this distance. After two long, hard years, I graduated with an excellent grade-point average and a degree I am proud of.

Since graduation, I have celebrated the birth of my second daughter, Jennifer, and I have progressed from an intensive care nurse to a flight nurse at St. Mary's Medical Center in Saginaw to director of emergency services at Central Michigan Community Hospital in Mt. Pleasant, MI.

MMCC has not only been an excellent source for my education, it has been a building block for my career. I have shared my experience and education as an instructor for emergency medical technicians, fire fighters, police agencies, the American Red Cross, and my fellow nurses.

My position affords me the opportunity to continue my education. I plan to attend Central Michigan University and obtain a bachelor's degree in business administration with a minor in health care administration.

Stephen M. Kalafut graduated from Mid Michigan Community College in 1986 with an associate's degree in nursing. After holding the positions of intensive care nurse and flight nurse, he currently serves as director of emergency services at Central Michigan Community Hospital in Mt. Pleasant, MI. He has shared his experience and education as instructor for EMTs, fire fighters, police agencies, the American Red Cross, and his fellow nurses.

Carol King
Gainesville College, GA

When Charlene Hunter and Hamilton Holmes became the first Black full-time students to attend the University of Georgia amid the sound and fury of angry mobs and armed escorts, I silently vowed, "I'm going to go to that school." Despite my 10-year-old optimistic idealism, I knew fulfilling that vow would not be easy for a shy little Black girl from a low-income family in a small Southern town of 1960. In less than 10 years, I began realization of that childhood fantasy by entering Gainesville College (formerly Gainesville Junior College).

Funny thing, though, the path to my goal actually was a detour. I'd chosen to sacrifice my original plans of attending a historically Black college in south Georgia for the love of my high school sweetheart. Shortly after graduation in 1967, we decided to get married and start a family.

Attending college seemed like a faraway dream for a wife and mother who had neither the time nor funds to make it come true. But a family friend encouraged me to try GC. I could attend day classes and work evenings and weekends to cover tuition and other costs, or maybe even qualify for financial aid.

My first quarter, winter 1969, was tough: juggling Math 101 and child care, working in the college library and keeping house, and, most challenging, being student, mother, and wife. By the end of my second quarter, the marriage lost.

A chance meeting of the College Players drama club changed my life's direction. The drama instructor had developed a readers' theater production for a national competition. A Cuban refugee with a sensitivity for the problems of minorities, he focused the production on discrimination and racial pride. When I became one of the four cast members, winning first place trophies helped me realize the versatility of my abilities. My self-esteem skyrocketed, especially after winning top honors for a two-person performance at another competition.

With this new-found confidence, I re-evaluated my career goals. I had always been interested in written and oral communications, but I'd never considered pursuing a communications career until my adviser, a graduate of the Henry W. Grady School of Journalism, talked of attending classes with Charlene Hunter. I decided to follow the path to journalism. In fall 1970 I completed studies for an associate degree in journalism at GC. By summer '72, I earned a bachelor's degree in journalism from the University of Georgia.

Fulfilling that childhood fantasy was only the beginning. I launched my journalism career at my hometown daily newspaper, where I believe my minority perspective contributed to the balance of news coverage. I frequently talk with and encourage young people to pursue higher education and to consider starting at a two-year community college.

Carol King graduated from Gainesville College in 1970 with an associate's degree in journalism. She now lives in Atlanta, GA, where she serves as director of public relations for Northside Hospital. Ms. King has been active in civic and cultural organizations, including serving as a board member and secretary of the Atlanta Press Club and serving four years on the Board of Governors of the Arts Council of Gainesville.

Lauren Gail Kyber
Piedmont Virginia Community College, VA

After graduation from high school 20-some years ago, I entered college with the full expectation of completing the four years required to earn a degree in fine arts. However, family circumstances forced me to drop out in my sophomore year. I began working as a secretary for the local department of social services, and somehow the years went by. A marriage and two children occupied my life for awhile. When the marriage ended in divorce, I realized that I was almost 40 years old and I was still working in a profession that was never my first choice and was far from fulfilling.

While I had always chosen to work in human services, as a secretary I always felt one step removed from the people being served. I really wanted to be working one-on-one in a helping profession. I felt trapped, believing there was no way for me to go back to school and still meet my responsibilities at home. One fateful day, I saw an announcement in the paper of registration dates for Piedmont Virginia Community College. One of the degrees offered was an associate degree in nursing.

I took a risk and called for an appointment at PVCC. Their counselors and financial aid officers helped to put it all together for me, and somehow everything fell into place. I began what was to be the most challenging, exciting, and fulfilling venture of my life just two weeks later.

I graduated six months ago and am currently working as a registered nurse in a rehabilitation center for children. My long-term goal is to work in home health care where I can develop a one-on-one relationship with each patient. Eventually, when my children are grown, I would like to be a nurse with the American Red Cross or the Peace Corps, which would offer me the opportunity for travel while helping people in other parts of the world to meet their health care needs.

I am so very grateful to all those at PVCC who encouraged and assisted me in reaching my goals.

Lauren Gail Kyber graduated from Piedmont Virginia Community College in 1989 with an associate's degree in nursing. She now lives in Charlottesville, VA. As a registered nurse at Kluge Children's Rehabilitation Center, she serves thousands of children each year. Eventually, when her children are grown, she would like to be a nurse with the American Red Cross or the Peace Corps.

June M. Laing
Roane State Community College, TN

Eleven years ago, at age 40, I found myself a single parent with four minor children, one emancipated daughter, and no marketable skills. For the first time in a long while I had to set long-range goals, the most important of which was to earn another degree, which would provide for me in the future.

Roane State Community College is in a pastoral setting, but make no mistake, it is as demanding as the prestigious southern university from which I later received my bachelor of medical science degree. I remember my first day at the RSCC bookstore when I handed back the Johns Hopkins Atlas of Human Functional Anatomy and said, "Oh, I don't get this. I'm not pre-med. I'm in medical records." The attendant (speaking lines straight from the sergeant in "Private Benjamin") said, "YOU GET THIS." Fear began its numbing process at my toes and worked upward. What had I gotten myself into?

Three years later, it was apparent that what I had gotten myself into was a deep love affair with medicine and computer science. I studied under a group of teachers who were the embodiment of the word "teacher." I felt an obligation to do the best work possible in response to their commitment to teaching. Perhaps part of it was my trying to learn all I could, knowing I would never again have the opportunity to take four years of my life to study. But I so appreciated all the intellectual doors that were opening to me. To reveal what?—all that I didn't know. And isn't that the ultimate lesson?

With the thorough and strong foundation in medical records technology laid at RSCC, I next completed a paralegal course and courses for my BMS. After working in a hospital for about a year, I began my consulting career. It was my turn to teach. Once, a group of students thanked me and said I had taught them so much. I told them that whether I had or not, I hoped that I had made them curious. That would last longer.

Since then I have had client hospitals from Oregon to Florida and Vermont to California; Panama; Iceland; the Azores; Belgium; Spain; and Germany, where I lived. Two years ago I became the manager of the DRG Support Services for Humana, Inc., based at corporate headquarters in Louisville. I manage a team of DRG consultants who review medical records and coordinate annual educational seminars for our 82 hospitals. During the past year I have been developing a computer-based education program that will teach medical terminology, anatomy/physiology/pathology, and medical record coding to personnel in the company and our hospitals. And where did I go for collaborators to write the courses? Back to those who taught me at RSCC.

June M. Laing graduated from Roane State Community College in 1981 with a degree in medical record technology. She now lives in Louisville, KY, and has gained extraordinary success in her career since graduation from RSCC. She is recognized throughout the country as an expert in the field of medical record coding. In her position as manager of DRG Support Services for Humana, Inc., she is currently developing a computer-based education program that will teach medical terminology, anatomy/physiology/pathology, and medical record coding.

Carol L. Lang
Mid Michigan Community College, MI

I have become a physician, which for many years was supposed to be an impossible dream. After all, I was a farmer's daughter from a small high school, and no one in the immediate family had earned a college degree. Community college made it financially and emotionally easier to make the transition to a university and then to medical school. It gave me time to really define what I could do scholastically to develop confidence in my abilities. Also, it provided the opportunity to find out who I was and whether I really liked that person, without the heavy pressure of being just a number in a large university setting. I have found over the years that many of the supposedly physical and intellectual barriers became surpassable if I just tried—sometimes repeatedly. Working hard for what I finally determined I wanted to do was definitely worth the effort.

My studies at Mid Michigan Community College, 1969-1971, prepared me to do as well at Central Michigan University as I had at MMCC. In my senior year, I determined that secondary education was not my desired goal. Osteopathic medicine was. During the next one-and-a-half years I completed my pre-med prerequisites and some master's degree work and was accepted to the Osteopathic College of Medicine and Surgery in Des Moines, IA. Though apprehensive, I embarked on the medical school program because I knew I had learned how to ask for help when I needed it. I graduated in 1979. After interning in Milwaukee, WI, I decided that general practice, which is more teaching than I realized, was my ultimate goal.

I came to Colorado and obtained my own private practice in 1981. My main goals as a teenager and a college student have been realized or replaced with new ones. I know my life would not have been the same without the professors and small college environment providing a good foundation upon which to build future success. It has become very important to me in my practice to try to explain, in terms patients can understand, what is wrong with them and the ways medicine can help change their lives, defusing their fears. Understanding is the greatest and best tool I have to help patients deal with disease.

Carol L. Lang graduated from Mid Michigan Commuity College in 1971 with an associate in science degree. She now lives in Arvada, CO, where she operates her own private practice as an osteopathic physician. After graduating from the Osteopathic Clinic of Medicine and Surgery in Des Moines, IA, in 1979 she began serving as clinic director of Monbello Clinic at Rocky Mountain Hospital.

J. David Martin and Barbara B. Martin
Southeastern Community College, NC

In the fall of 1986, after years of operating a successful business, my wife Barbara and I were faced with an important decision: "What should we do with the rest of our lives?" We were both 35, and I was recovering from a long illness and brain surgery. Thinking of how fortunate we were that I would fully recover, we wanted to repay society. I decided to go into nursing in order to give back some of the professional care I had received. I knew that starting a nursing program while I was still recovering would be tough, but my wife said, "Let's do it together, just like we've always done." This was a huge challenge for her, since she had only completed the ninth grade. She began studying at the learning center at Southeastern Community College and got her high school equivalency certificate. After passing SCC's nursing admissions tests, we were ready to take prerequisite courses and enter the associate degree nursing program.

We made it through the three years with encouragement from our family, especially our 10-year-old son, and with a lot of guidance and inspiration from our instructors. In a candlelight ceremony on May 28, 1989, we obtained our goal when we were pinned as ADN graduates.

Throughout our training, Southeastern met our educational and psychosocial needs and provided continuous moral support. We were not surprised in the least to pass our N.C. Board of Nursing exams on the first try!

As registered nurses, my wife and I now work with medical and critical-care patients at a community hospital that emphasizes a caring atmosphere. We are able to look back now and really see what SCC was able to do for us. Barbara says, "I have self-respect and self-confidence that I never had before enrolling at SCC. I've gone from a ninth-grade drop-out to a college-graduate RN!" We'll forever owe our thanks and appreciation to the people at SCC for making our dream come true.

J. David and Barbara B. Martin both graduated from Southeastern Community College in 1989 with associate's degrees in nursing. They reside in Lake Waccamaw, NC, and are employed as registered nurses at Bladen County Hospital. They are older students who made mid-life career changes. David had earned a BS in business administration in 1978; Barbara had never finished high school. At SCC, they completed the associate degree nursing program together and became RNs.

Evon Miller
Kaskaskia College, IL

I was married shortly after graduation from Centralia High School. Three months after our second anniversary, our first child was born. Further education at this time did not appear to be in my immediate future.

After the birth of our second son, I became interested in nursing. I enrolled in a local LPN school, graduating in 1967. I obtained a job in a prominent doctor's office in Centralia and remained employed there for five years.

By this time our family had grown to four. My husband was attending Kaskaskia College part-time while working full-time. He would come home enthused and excited about his classes and professors. His excitement and enthusiasm were infectious. With my husband's encouragement, I also enrolled in Kaskaskia College, majoring in nursing.

Looking back, I remember all my doubts, fears, and insecurities those first days on campus. I thought Kaskaskia College would be cold, impersonal, formal. It was a pleasant surprise when I realized that the instructors really cared about me and really were concerned about my progress.

Enrolling in the RN program was the most challenging and rewarding act of my life. This program not only prepared me as a registered nurse, but also instilled self-confidence.

My husband and I graduated from Kaskaskia College in 1974. Since that time, I have been employed at Warren G. Murray Development Center in Centralia. My husband continued to SIU-Carbondale, obtaining his BA in 1975.

Now that our youngest son is enrolled at Kaskaskia, I have begun working toward my BSN. I realize that none of this would have been possible without the help and encouragement I received at Kaskaskia College.

Evon Miller graduated from Kaskaskia College in 1974 with a degree in nursing. She now lives in Centralia, IL and works with the developmentally disabled children who reside at Warren G. Murray Development Center. She is effectively using her registered nursing skills obtained at Kaskaskia College within the area to provide appropriate care for individuals who cannot care for themselves.

Margaret Mose
Johnston Community College, NC

In 1976 I was a single parent with four dependent children. I had married very young, and I had few job skills. I was working as a domestic servant in a home in Waterbury, CT, and my future looked bleak.

As a girl, I had dreams of being a nurse. I'd always see myself in these dreams being transformed like Superman into a nurse, able to help everybody. But with four children to support, going to school to become a nurse seemed impossible.

But I attended a revival in Waterbury, and the spirit of the Lord began to deal with me and to speak to me. The Lord clearly said "Move!" so I packed my kids in our beat-up old car and drove to Raleigh, NC.

I took a job as a maid in a Raleigh motel at $58 a week. Those were rough days for the children and for me. Friends and fellow church members kept us from going under. One winter we couldn't pay the gas bill, and was our house ever cold!

At one time I was forced to go on welfare for a while, but I got off the food stamps as soon as I could. The Lord always provided.

In a few months I began to work at Wake Medical Center in Raleigh as a nurse's aide. In 1978 I was working for $2.76 an hour. Sometimes I had to hitchhike to work.

In 1984 I entered the nursing program at Wake Technical College, but financial pressures, family demands, transportation problems, and a heavy work schedule were too much for me. I failed several courses and had to drop out.

But I still had that dream of being a registered nurse. I entered the nursing program at Johnston Community College in Smithfield, more than 20 miles from my home in Apex. I still had many transportation and financial problems, but I had teachers who would drive me to classes, as well as classmates and church friends who helped out with food and other necessities. Johnston Community College provided financial assistance, and I began to feel that I really might make it.

In August 1988 my JCC classmates and I took the North Carolina State Board of Nursing exam, and 100 percent of us passed. When I saw that "RN" by my name, you can believe I praised the Lord!

At present I am a fully certified RN working in the operating room at Wake Medical Center. I am working at an important job where lives are saved every day, and I have realized my dream.

I am proud to be a graduate of Johnston Community College, where the faculty and the staff give students encouragement and support from start to finish. There are many schools that would have considered me a poor risk, but JCC believed in me and gave me the strength to acquire the training I needed.

> *Margaret Mose graduated from Johnston Community College in 1988 with a degree in nursing. She now lives in Apex, NC, and works as an operating room nurse at Wake Medical Center in Raleigh. She had to overcome staggering personal obstacles in her quest to realize her girlhood dream of becoming a nurse. Patients in her care instantly sense that she is a strong person who can help them get well.*

Francisco Rivas
Miami-Dade Community College, FL

I graduated from high school in 1968 and then entered Miami-Dade Community College. After not doing well for a couple of semesters, I decided to enroll in a technical school, where I graduated in 1972. Feeling that I was not getting anywhere in the job market, I made the decision of returning to M-DCC. By this time I was married and had to work over 40 hours a week to maintain my home and pay for school. With a lot of encouragement from my wife and my teachers at school, I was able to feel very good about my progress and did well in my classes. I graduated from Miami-Dade with an associate of science degree in medical technology and a 3.7 GPA. For the first time I felt prepared to get a good job, and thus continued educationally to get a higher degree. A few years later I received a bachelor's degree in health services administration, graduating with a 3.9 GPA. I am presently pursuing a master's degree in the health services field.

What started at Miami-Dade Community College in 1972 has allowed me to become a college teacher for over 10 years, supervisor of a microbiology department of a large hospital in the Miami area, and assistant administrator of a chain of medical centers in Dade County, FL.

What I cherish the most is that some of my years of teaching have been at M-DCC, allowing me to give something back to a place to which I owe so much.

Francisco Rivas graduated from Miami-Dade Community College in 1976 with an AS degree in medical technology. He now lives in Miami, FL, where he is an assistant administrator of the Palmetto Medical Centers. Since receiving a bachelor's degree in health services administration, he has returned to Miami-Dade Community College, Medical Center Campus, to help students review for their state board exams and to serve as chair of the Medical Laboratory Technology Advisory Committee.

Jo-Ann Rossitto
San Diego City College, CA

When I decided to enter the field of nursing, I was extremely frightened and uncertain. I had no idea what nursing was all about. I still was struggling with "What am I going to be when I grow up?" The only "given" in my life (at 18 years of age) was the intense desire to help people—to give of myself for the betterment of mankind.

My journey began with the vocational nursing program in the San Diego Community College District, both at City College and Mesa College. The experience was to shape my destiny and give meaning to my existence in a career where another person's life rests in the palms of your hands, where an inaccurate decision may mean the difference between life and death.

I remember the extreme dedication and patience of the instructors who guided me through an intense educational process—one to which only military inductees could relate! Some were friendly, others intimidating, but all were inspiring.

I learned from them to pursue my dreams and believe in myself; to search beyond the obvious; to listen; and most of all, to respect and love the people to whose well-being I would dedicate myself.

In 1970 I passed the California State Vocational Nurse licensing exam, then worked with the elderly in a convalescent hospital. Yearning for growth, I enrolled in the community college LVN-to-RN program, and in 1972 I received the coveted title of registered nurse. Wow! I had made it . . . I was a real nurse!

I loved my first RN job at the Veterans Administration Medical Center, and decided to earn my bachelor's of science in nursing degree at the University of San Diego. I expanded my experience into other hospital departments, and also to home-care nursing.

Yet I found another desire—to teach. In 1981 I earned a master of arts in nursing education at New York University. Returning to San Diego, I began teaching nursing at my old alma mater, San Diego City College. My new colleagues had once been my instructors!

It was my ultimate desire: to be for my students what others had been for me, years earlier. In a few short years, I faced yet another decision. The director of the nursing program was retiring, and the position was open. Should I apply?

Yes, I applied, and I was appointed associate dean and director of nursing education at City College. I had made it to the top of the professional ladder.

I continue to campaign for student success and excellence in education. We must enable others to move from novice to practitioner in any chosen field. By my success, I am a prime example of that philosophy.

Jo-Ann Rossitto graduated from San Diego City College in 1972 with a degree in nursing. She now lives in San Diego, CA, and is currently the associate dean and director of nursing education at San Diego City College. She earned a master of arts in nursing education at New York University in 1981 and returned to San Diego to teach nursing at her alma mater. It was her ultimate desire to be for her students what others had been for her years earlier.

Aristeo H. Soto, Jr.
Central Arizona College, AZ

Coming from a rather large family (four brothers and six sisters), I found out early I had to work hard for what I got. Being a farm laborer and the sole provider, my father had an extremely tough job providing for our family. It was because of this that many of us had to help out by working summers and after school throughout our school years. Working out in the cotton fields during an Arizona summer makes one appreciate the value of a good education. It was during my high school years that I knew I wanted to achieve something and better myself academically, but I had not really anticipated going on to college because my family just couldn't afford it.

It was my scholastic and athletic abilities that enabled me to pursue a college education. Enrolling at Central Arizona College was a major step for this naive but determined farm boy. Being the first in my family to go to college made me that much more determined to succeed and make my parents proud.

After completing one year of studies at Central Arizona College, I was still uncertain in which direction I was headed. I enlisted for four years in the U.S. Air Force and then resumed my studies at CAC. I had always been concerned about my own health, but it wasn't until I enrolled in a human anatomy and physiology class that I realized I wanted to help others achieve and maintain optimum health. The workings of the human body in chiropractic fascinated me, and I believe it was this one class that motivated me to pursue a career.

After completing my studies at Central Arizona College, I continued at Arizona State University where I graduated in 1978 with a BS in physical education. I graduated from Palmer College of Chiropractic in 1986.

I have found great joy in helping others through chiropractic.

Central Arizona College . . . thank you for the guidance and encouragement given to me even after my studies ended at your college.

Aristeo H. Soto, Jr. graduated from Central Arizona College and currently resides in Casa Grande, AZ, where he is a chiropractic physician. It was his studies at CAC that motivated him to pursue a career in this field. After completing his studies at CAC, he continued at Arizona State University and then later graduated from Palmer College of Chiropractic in 1986. As a chiropractor, he helps people in his community live a pain-free life using natural healing methods.

Rosemary T. Stewart
Delaware County Community College, PA

After high school, I had limited finances as well as doubts about my ability to meet the challenge of a four-year institution. I decided to apply to Delaware County Community College, where I could find out if I had the ability to develop the discipline to study and to clarify my personal career goals. The counselors at the college were instrumental in procuring financial aid, which enabled me to concentrate on my studies. As a work-study student, I gained my first experience in library research, which was to become a valuable asset in future endeavors.

The following year I took a work-study position in the science laboratory. This was to be a turning point in my choice of career. While preparing the laboratory assignments I discovered a proficiency and interest in health science studies. At this time I decided to pursue a career in a health care laboratory. I reinforced my interest in coursework in anatomy and physiology and chemistry. I applied and was accepted into the School of Histotechnology at the Hospital of the University of Pennsylvania. I also continued my formal education at Neumann College, where I received a bachelor of science in 1981. I have subsequently been promoted to research assistant in the Pigmented Lesion Group of the Hospital of the University of Pennsylvania.

Because of my interest in my chosen area of research (skin cancer) and my desire to increase public awareness about this dreaded disease, I decided to pursue a master's degree in health care education. I graduated second in my class from Saint Joseph's University in May 1989, and I am now enrolled in a post-master's training and development program.

Delaware County Community College provided me with the opportunity to test the demands of higher education. I discovered the value of discipline and perseverance. The self-confidence and sense of responsibility earned in the science department provided the foundation for my career development and will always be valued. The comradeship I enjoyed during my participation in the community chorus concerts with my peers will be remembered forever.

The task that is presented to the faculty of the community college is challenging: to provide a first-rate formal education to students and a source of practical knowledge for the community. If my own experience is a reflection of their performance, they deserve life-long thanks for a job well done.

Rosemary T. Stewart graduated from Delaware County Community College in 1975 with an associate's degree in science. She is now living in Upper Darby, PA, and works as a research assistant in skin cancer in the Pigmented Lesion Group of the Hospital of the University of Pennsylvania. Because of her special concern for preventing skin cancer, she gives talks and takes part in educational programs to disseminate her knowledge on the topic.

H. Katheryn Tatum
Front Range Community College, CO

I had wanted to be a nurse for as long as I could remember. When I was in high school, I tried to get one of the scholarships that were available to students, but I was not able to get one. I finally enrolled in a hospital program for practical nurse training. I graduated and took the state boards and went to work as a practical nurse. My dream had always been to come to Colorado and to be an RN. In 1960 I came to Colorado with my five children and went to work as an LPN. I still had not realized my dream of becoming an RN. I enrolled in a diploma school of nursing, and before the first year was over I became pregnant with my sixth child. I was told by the director that I needed to stay home and take care of my child. Since I had only missed one week of class, I felt that this was an unfair assumption.

I continued, however, to take classes at the University of Colorado at Denver. The nursing class was a five-year course. I had applied, but had not been accepted, when Front Range Community College opened. I received some information about the college and decided to look into it. I was very excited when I learned that they had a two-year nursing program. I applied for entrance, and the most exciting day of my life was the day I received the letter of acceptance.

The first class I took during the summer was chemistry. The instructor was interesting and very interested in the students. I was hooked when I made an "A" in both semesters. I knew I could make the grade. All I had to do was study. There was a class to teach students how to study and how to take tests. After that, I took off running and never looked back. My family was as excited as I at graduation in May 1972. Front Range Community College will always be my college roots. It was where I got my start. It was geared for students with families as well as the traditional college student.

I went on to transfer my credits to a four-year college and finished my BSN in 1974. I started to work at the University of Colorado Dialysis Unit as head nurse. I realized that I needed more educational tools to be good at my job. I applied to graduate school, was accepted, and graduated in two years with a master's of science in nursing in 1978. Front Range Community College is the reason that I am where I am today. I thank God that it was there when I needed it.

H. Katheryn Tatum graduated from Front Range Community College in 1972 with an associate's degree in nursing. She now lives in Aurora, CO, and is the unit administrator of the Adult Acute Dialysis Unit at the University Hospital in Denver. In her community, she coordinates health fairs, conducts in-service training for public schools and health care agencies, speaks to nursing students on career mobility, and tutors students in problem-solving techniques. Additionally, she works on projects to promote literacy and help battered women.

Paul D. Taylor
Garden City Community College, KS

The world was simple and clear cut in 1955. I did not make plans for life after high school. In fact, I retreated into a sort of fantasyland where working was for the sole purpose of maintaining a "cherry" 1940 Ford, dragging main street, and cruising for chicks. Life was great.

The week before Labor Day, my world began to come apart. My girlfriend would soon go away to college and my buddies were preparing for the same. My employment was ending and my father was on my case for a perceived lack of motivation, direction, and a sense of the need to do something worthy with my life. I listened, sulked, and wallowed in self-pity. Emotional goodbyes were followed by the strange urge to give college a try. After all, I didn't have anything else to do, and the college was just six blocks away.

The community college became the focus of my life. I even began to hit the books. I was not viewed as a brain in high school, but this was different. I began a love affair with my subjects and my instructors. I honestly passed courses, looked forward to the next semester, and completely emerged from the pit of perceived societal oppression, discrimination, minority expectations, and limited opportunites.

After graduation, I attended the University of Kansas for one year. I did not return the next fall and did not graduate. Instead, I took a bride and moved to Colorado. I found employment on a newly constructed research team. Our goal was to perform organ transplants in humans. This was a brand-new field. My contribution, supported by my community college education, acquired work ethic, and the input of others, culminated in our team performing the world's first human liver transplant in March 1963. These pioneering transplant procedures are now applied for the benefit of mankind throughout the world.

My professional successes include my inclusion in national registries such as *Who's Who Among Black Americans*; publication in medical journals, books, and magazines; many television and radio appearances; and world travels for lectures, consultations, and the initiation of new transplant programs.

When my last child graduated from college and started law school, I finally understood that regardless of the changes that have occurred in society during my generation, the interest and drive that I found during my community college education motivated me to motivate my children...and that's good for America.

The fire that started burning in my gut that year still burns today. I learned something about slow starters, love and gentleness, respect for constituted authority, and the increasing, important role of the community college in the educational goals of America.

Paul D. Taylor graduated from Garden City Community College in 1957 with a degree in education and now lives in Pittsburgh, PA. He is senior instructor in surgery at the University of Pittsburgh and has become the world's senior transplant coordinator. Author of a manual on the procurement and preservation of cadaver organs for human transplantation, he is the template from which the world's transplant coordinators have been struck. Professional successes include publications in medical journals, many television and radio appearances, and world travel for lectures and consultation.

Mary Anne Vaughn

Southwest Missouri State University-West Plains Campus, MO

Several years after graduation from high school, after years of drifting through the job market (as a cook and waitress, sewing garments and shoes, even hauling cord wood) I decided something had to change.

Economic obstacles stopped me from attending college right out of high school. I had no skills or professional abilities to help me overcome being chronically unemployed and underemployed. Education was the only way for me to get ahead.

In 1979 I moved to St. Louis, attended one year of vocational training, and became a licensed practical nurse. As soon as I began working, I realized nursing was for me. After a year or so I decided I should be an RN. I was learning rapidly from my nursing experiences and quickly outgrowing my title and the professional restrictions of my LPN.

Southwest Missouri State University-West Plains was at that time beginning its associate degree nursing program. In 1983 I began taking chemistry, microbiology, and English, chipping away at degree requirements, waiting anxiously to be admitted to one of the first few nursing classes. When that day finally came, I knew I was opening many doors and closing some behind me forever.

I graduated in 1986 with an associate's degree in nursing and subsequently was employed by Ozark Medical Center in West Plains. Today, I am head nurse of the dialysis department at the hospital.

The moral of my story is that education is a process; it's done step by step. Vocational training and associate degree programs, together with life experience, can be important components of that process. As the process of education continues, the door is always open to new challenges. SMSU-West Plains and the education I received there provided my door of opportunity.

Mary Anne Vaughn graduated from Southwest Missouri State University's two-year campus at West Plains in 1986 with an associate's degree in nursing. She now lives in West Plains, MO, and is a head nurse in the Dialysis Clinic at West Plains Ozark Medical Center. She has become a key employee in the hospital's dialysis clinic and is especially appreciated by the large number of elderly patients who rely on the health care facilities in West Plains for medical needs.

Gail Winfrey Winston
DeKalb College, GA

By the time I applied to DeKalb College (formerly DeKalb Community College), I had been out of high school for four years and had attended three different colleges. My parents jokingly accused me of collecting college IDs; I just had no idea what I wanted academically or where I was headed concerning a career. College simply represented my ticket to experience life—to be independent, to set my own limits, to create my own lifestyle, and, most of all, to have a good time. I took courses that sounded fun and studied just enough to get by.

Then, thanks to a minor personal crisis, I awoke to the fact that the generous financial pipeline to my parents was not a permanent fixture. I might have felt independent, but as long as I was being financially supported it was a false freedom. With an incredible array of emotions ranging from nonchalance to anxiety, I applied to and was accepted into the nursing program at DeKalb. The nonchalant side of me said I could quit the program any time it became "un-fun;" the anxious side of me worried about actually making a year-long, goal-oriented commitment. My last declared major had been Eastern religions; did I even want to be nurse, anyway?

Those two short years provided the most dramatic turning point in my life. It remained important for life to be fun and exciting, but there was now also room for seriousness, discipline, and academic achievement. It became my choice to study and to succeed, and the acquisition of knowledge and skills provided a new-found, intense satisfaction. My instructors were committed, vibrant role models who instilled in me a respect for academia in general and nursing in particular that I had not previously experienced.

When I graduated from DeKalb, I knew this was not the end of school for me. After working for a year as an emergency department nurse, I entered a BSN program. After another few years of emergency nursing I entered a graduate program in nursing and earned a master's degree in adult health nursing. The success I experienced in the latter two programs was a direct result of the quality of education I received at DeKalb. The success I have experienced in my profession reflects the high standards, commitment, and caring of DeKalb's instructors.

What began as a quest for financial independence turned into a love affair with learning. My worries about nursing being "un-fun" evolved into a commitment to a profession that has consistently offered me continuing growth and satisfaction. All this because of DeKalb College. My heartfelt thanks.

Gail Winfrey Winston graduated from DeKalb College in 1976 with an associate's degree in nursing. She now lives in Decatur, GA, and works as a coordinator of lifestyle and health education. She was an emergency department nurse for 10 years. In her current position she develops, writes, and teaches wellness programs including stress management, smoking cessation, PMS education, comprehensive first aid, women in transition, and AIDS education.

CHAPTER SEVEN

Social Services

Wilbur T. Aldridge
Rockland Community College, NY

When I had completed my high school education in North Carolina, it was my family's desire that I attend college. So I did. Without giving it much thought, I enrolled in a large university. Immediately, there was a sense of estrangement. I felt like a number among many numbers. For the first time, I was with strangers, all of whom were vying for positions and friends. I felt alone and pressured and left the university after one year.

The fact that I left college meant I had to have a job. Opportunities were scarce in North Carolina, so I left and found employment at a residential facility for the developmentally disabled in Rockland County, New York. It was not long before it became apparent that my career choices were extremely limited without a college degree, and advancement was virtually non-existent. It was then that I decided I must return to college. Fear of failure almost undermined my decision, but the desire to advance was much greater. I knew I had to take the risk, and I chose Rockland Community College. Like most students, I had to tackle a battery of placement tests. One of the tests indicated my interest in people, so I was guided into majoring in human services. The Human Services Department at Rockland Community College consisted of about 100 students and six professors. This time, I was not a number but a real person interacting with other people, including the faculty. The individual attention from faculty and counselors was tremendous. Their obvious belief in my ability to achieve provided me with the confidence to believe in myself.

It was this belief in myself that enabled me to complete my associate degree requirements and immediately enter a four-year university. The end result was my completion of the requirements for a bachelor's degree in social work.

I owe Rockland Community College a great deal for instilling in me self-discipline and for providing me with a positive self-image, thereby giving me the ability to continue my self-development. These are the tools I proudly acquired through the encouragement, individual space and respect, and the nurturing afforded me by the professional staff at the college.

My college experience also provided me with a sense of "don't forget home," a belief that one must continue to serve the college and the community at large. I have tried to live by this philosophy. In doing so, I became involved with community organizations including the Rockland Community College Alumni Association, of which I am proud to serve as president.

From a "go nowhere" job, my education has enabled me to advance to the position of executive assistant to the director of Letchworth Developmental Disabilities Office for the Developmentally Disabled.

Wilbur T. Aldridge graduated from Rockland Community College in 1971 with a degree in human services. He resides in Garnerville, NY, where he serves as executive assistant to the director of Letchworth Developmental Disabilities Office for the Developmentally Disabled. He has made substantial contributions to the lives of developmentally disabled persons and those who care for them. He serves on the boards of directors of such community service organizations as the Rockland County Girl Scouts and the Thiells Child Care Center.

Mary Bryant
Tulsa Junior College, OK

Today is someday! That stark realization drove me, a woman then well past 55, to enroll in English Composition I and American Literature I at Tulsa Junior College in 1984. I told friends that I wanted to see if I could still study. I hadn't mentioned the tiny flame of desire for a college degree that I secretly nurtured. I remembered the notes several English teachers had written many years before, suggesting that I develop my writing skills. On the day I enrolled in 1984, I knew that today was someday.

I can only describe the emotion I felt as I entered that first classroom as cold, icy fear. Almost 40 years had elapsed since my last classroom experience. By the time class ended, I was caught up in the lecture, the questions and comments of other students, and my own interest in the subject. I felt the flame of my long latent desire burn brighter.

To my surprise, my professors and fellow students did not treat me as an oddity. Before the semester had ended, these understanding instructors, who have since become my mentors, helped me select courses for the following semester. Each new course opened new vistas and stimulated new interests. I felt a growing sense of accomplishment and self-confidence.

Whenever the old fear returned, and it did, I sought out my mentors for understanding and encouragement. We began to discuss an associate degree in English, which I was awarded in 1987. I expect to complete my BA in English from Northeastern State University in 1990.

When I have my degree, I plan to teach English or creative writing on a part-time or volunteer basis. I am presently involved in the GED literacy program at the Tulsa Public Library. One day a week I work as an English teacher's aide in that program at the county jail. Because of the skills I began to develop at TJC in 1984, I am confident that I can make a significant contribution to my community. I can make a difference.

Tulsa Junior College is a community treasure. It is a beacon to people from all walks of life. It offers to graduating high school seniors a firm foundation for continuing their education in a university or earning an associate degree in a variety of two-year programs. There are fine facilities for the handicapped and evening classes and short courses for those employed full-time. Best of all is the warm welcome extended to returning students of all ages. To all who have said, "Someday I would like to continue my education," today is someday.

Mary Bryant graduated from Tulsa Junior College in 1987 with a degree in English. Her home is Tulsa, OK. Her community service activities include acting as a literacy volunteer at the Tulsa Public Library and in the county jail. A creative writer, she has dedicated herself to teaching English and creative writing on a part-time or volunteer basis once she earns the bachelor's degree that she is now pursuing.

134

Manuel Chavallo
Columbia Basin College, WA

I graduated from Kennewick High School in 1978 with a poor grade point average. During my senior year in high school I had still not applied to college and did not have any strong desire to do so. Even though my "C-" average in high school was somewhat weak, it was a great achievement for me because I was the fifth child of nine and I was the first in my family to earn a high school diploma. Many times I would miss school in the springtime because I, along with my brothers and sisters, would have to cut asparagus to make money. My mother was a single parent who spoke only Spanish and raised all the children.

Despite all this, in the fall after graduating from high school I enrolled at Columbia Basin College and decided that I needed to broaden my opportunities. I had a very difficult time with my studies, and at that point I realized that I had not prepared myself for higher education. I made the decision to leave school. I was accepted into the local plumbers' and steamfitters' apprenticeship program and worked for the next four years in the nuclear industry as a construction worker. My experiences in construction were very rewarding; however, my wife, Hilda, who wanted very much to attend college, convinced me that I should seek employment in the local job market while she attended school. She enrolled at Eastern Washington University in the fall of 1984 after completing her associate degree from Columbia Basin College.

Six weeks before college started, I made a very difficult decision to attempt college for the second time. Upon my acceptance and enrollment at Eastern Washington University, I enrolled in the university's student support services program and immediately began working on improving my academic weaknesses. After gaining confidence in my abilities, I graduated from Eastern Washington University in the spring of 1986 with a degree in economics and a minor in labor studies. During the summer of 1985, I attended six classes at Columbia Basin College in order to receive my associate degree. I then transferred my AA degree to Eastern Washington University, which helped tremendously in reducing the number of courses needed to graduate with a bachelor's degree. I was on a very tight time schedule to graduate from college because I, like a great number of other minority students, wanted very much to return and be near my family.

After taking a chance and returning to college, one of my most important accomplishments was developing leadership skills. During my senior year, I was elected president of Eastern Washington University's Hispanic Student Organization. Today, I take time from my busy schedule to speak to high school students in my community about the importance of higher education. I strongly encourage them to attend Columbia Basin College.

Manuel Chavallo graduated from Columbia Basin College in 1985 with an associate degree in general studies. He now lives in Kennewick, WA, where he directs the city's Upward Bound program. He frequently addresses high school student audiences on the value of higher education. He is a member of the Hispanic Professional Association and received a national TRIO Achiever Award in 1989.

Monda H. DeWeese
Hocking Technical College, OH

Although my position as the manager of a photography studio and camera shop afforded me the opportunity to communicate with the public, I felt the need to do so on a more powerful level. Like many others who decide on a career change, I was not willing to make the commitment to a four-year degree program. I went to Hocking Technical College to look into the counseling curriculum offered there. The corrections technology section contained the majority of the "helping profession" topics, though I was convinced I would never utilize the criminal justice courses in the workplace.

How wrong I was! During my first quarter, winter 1982, I was fascinated by the corrections field. Though my grandfather was a psychologist at a state juvenile detention facility and my father has worked at the Ohio Department of Youth Services for over 30 years, the idea of a career in criminal justice never seemed the path for me. The true impact of technical education on my career came from the quality of advising and instruction that I received. The "war stories," as our instructors used to call them, were real-life experiences that they would weave into our class discussion and could not be found in any textbook. The opportunity to tap into that type of knowledge is rare, and one that I treasure.

I enjoyed being active in many aspects of college life at Hocking Technical College. I was a student senate delegate and participated in Alpha Phi Sigma, the national criminal justice honor society. Remaining on the Alumni Advisory Board keeps me updated on the college's activities and growth.

Since beginning my criminal justice education at Hocking Technical College, I have continued taking courses there ever since. I earned a bachelor of criminal justice degree in 1986 and finished my master's in political science/public administration just this month. In my position as the executive director of a 42-bed, adult-male, community-based correctional facility, I most often draw upon the educational foundation that I received during my associate degree program in corrections at Hocking Technical College.

It gives me great pleasure to be able to teach a few evening courses to the corrections technology students at Hocking Technical College. I am now able to share with them my "war stories" of work life in the field of corrections. Every now and then, I notice the raised eyebrow, the quizzical look, and the sparkle in the eye of a student who has made "the connection," and I smile. Now, in a small way, I am helping repay Hocking Technical College for the quality education I received.

Monda H. DeWeese graduated from Hocking Technical College in 1983 with a degree in corrections. Now a resident of The Plains, OH, she serves as executive director of a 42-bed community-based correctional facility called Septa Center. In addition to contributing to correctional affairs at the local and state levels, she frequently teaches evening classes in corrections technology to students at her alma mater.

James Gardner
Columbia State Community College, TN

I cannot recall a time when attending college was not in my plans. Many times I heard my father say to my brothers and me, "Boys, I want you to have the opportunity for an education that I didn't have." During my high school years, I was encouraged by my teachers and my principal to attend college. At the time, however, the question for me was not "Would I attend college," but rather, "How?" Because I was the oldest of five children in a family that did not have the financial resources to send a child to college, the "how" was a very big question for me.

However, in 1966 the first of Tennessee's community colleges began operation—Columbia State Community College. After my high school graduation in the spring of 1967, I entered CSCC. Through scholarships, a small loan, and being able to live at home while attending college, the question of how I was to attend college was answered, at least for the first two years.

For many of my fellow students, CSCC was similar in size to the high schools they had attended. For me, CSCC was a good transition between a small high school and the University of Tennessee, which I would later attend to study pharmacy. And as I reflect on my experience as a college student, and later as a seminary student, I realize that I had some of the best teachers in my entire college experience at CSCC. The chemistry and physics that I had not understood well in high school came alive for me at CSCC. And I had not realized that biology could be interesting until I had botany and zoology at CSCC. I thoroughly enjoyed the political science and history courses that I took as electives. The teachers I had at CSCC were very good at what they did: teaching.

After 10 years of practicing pharmacy in an independent pharmacy, the calling to become a minister in the United Methodist Church took me back to academic pursuits. I entered Vanderbilt University Divinity School in the spring of 1983. Again I found that CSCC had prepared me well for that endeavor. Despite my resistance at the time, all those papers I had written in the English composition and literature courses at CSCC helped me to develop the skills necessary to do the work required in a master's degree program in theological education. I graduated from Vanderbilt University with a master's of divinity degree.

I feel very fortunate to have been able to enjoy the rewards and share the responsibilities of two professions. I believe both have contributed something to the good of others. For a poor farm boy to be able to see some of his dreams come true is indeed a blessing for which I am grateful. CSCC will always have a place in my heart because it started me down the road toward the fulfillment of some of these dreams.

The Reverend James Gardner graduated from Columbia State Community College in 1969 with a degree in prepharmacy and now lives in Columbia, TN. The minister of St. Luke's and Bigbyville United Methodist churches, he is a leader of United Givers' Fund for ministers and the Tennessee Pharmacy Association, as well as president of the Maury County Ministerial Association.

Marie Keegin
Frederick Community College, MD

Nine years after I graduated from high school, I was feeling unfocused. I was primarily an at-home mom but worked weekends as a hospital clerk. Although I had a vague desire to continue my education, many personal obstacles deterred me. My husband accused me of creating these barriers for myself because of my unsuccessful high school experience and my years away from the academic environment. He challenged me either to register at Frederick Community College or stop complaining. The anger from this accurate summary pushed me to register. Through tears and fears on that trip to register at FCC, I wondered about many things: what I would encounter at FCC; whether I could succeed as a student; whether I would have a rewarding career someday; and how I could balance all these unknowns with my family responsibilities.

When I arrived at FCC, the friendly person in the admissions office was a sharp contrast to the horror stories I had heard about registration at the big universities. This warm, encouraging atmosphere was also prevalent in my classes. The professor affirmed my efforts, discretely noted my deficiencies, and suggested readings or remediation. I credit the teachers at FCC with motivating and preparing me for further education and my career. Once I graduated, I was able to handle my coursework at the bachelor's and master's levels as well as my job responsibilities. Because of the nature of my job in employment and training, and FCC's ability to tailor its programs to meet the community's needs, my association with FCC continues.

As the director of the Job Training Agency in Frederick County, I have the opportunity to encourage many frightened women to overcome their obstacles and fears and to register at FCC. I know that once they experience FCC, they will begin to visualize their own success as I did.

Balancing school, work, and home remains a challenge. Although I am no longer primarily an at-home mom, my children remain my number-one priority. I was elated when my son decided to attend FCC. He was young, an average high school student, and a late bloomer. But I know FCC can do for him what it did for me.

I will always be grateful for the education and focus that I received at FCC. FCC continues to make an essential contribution to the development of our community and the education of its citizens. As the future life stories of my son and Job Training Agency students at FCC are told, FCC will undoubtedly have played an important role in each of their lives as it has in mine.

Marie Keegin graduated from Frederick Community College in 1980 with a degree in business administration. She now lives in Frederick, MD, and serves as director of the Frederick County Job Training Agency. Through her work at this agency, she is also able to encourage numerous men and women, many of whom are uncertain of their own talents and abilities, to take positive steps toward fulfilling their potential.

Rose Mary Lucci Kirwin
Aquinas Junior College at Milton, MA

My blue-collar family expected me to become their first graduate of a four-year college. This would be for them a dream come true and a sign of their success in the "new" world. While I was in high school, my father, a machinist, was involved in long and costly strikes. During my senior year, he told me that he couldn't pay for my college education unless he took a second mortgage on our home. I couldn't allow this, because our home was all we had. So, I put aside my dreams and applied to Aquinas Junior College, explaining my financial predicament. The college allowed me to enter school and repay my obligations when I graduated.

As September approached, I dreaded entering school—no longer was I going to "discover a vaccine to cure the ills of the world," but I was going to learn to be a secretary! This attitude dissipated almost as soon as I arrived. I discovered Aquinas wasn't made up of students who had just taken the "business (easy) track" in high school, but those like myself who had been in college preparatory. This wasn't the "breeze" I had expected. But each time I faltered, someone was there encouraging me.

I realize now that Aquinas prepared me both for employment and for life in general. An old adage says: "If you give a man a fish, he is able to eat today, but if you TEACH him to fish, he will feed himself forever." This describes what Aquinas did for me.

Upon graduation, I was able to support myself, help my family, repay my school obligations, and obtain progressively more responsible positions. I am not afraid of learning new machines or systems, and my "old" skills will never go out of style. I was not afraid to help found a shelter for battered women, and I had enough faith in myself to move on to another challenge when my task was complete.

Today, as director of public relations and development in a non-profit agency, I use the skills I learned at Aquinas daily to raise funds for the disabled. At Aquinas, I was encouraged to excel and to develop my potential. I surpassed my own expectations. I believe now that if I had entered a four-year college from high school, I would have always been an average student. Aquinas helped me to mature and grow. I developed a thirst for learning, which I finally realized years later, graduating summa cum laude from a four-year college.

I believe that my Aquinas degree is invaluable. Aquinas gave me the tools for a meaningful and rewarding life—the understanding, joy, and fulfillment of a job well done.

Rose Mary Lucci Kirwin graduated from Aquinas Junior College at Milton in 1961 with an AS degree in secretarial science. Today she lives in Braintree, MA, where she serves as director of development and public relations for Center House, Inc. This facility is dedicated to improving the lives of emotionally disabled and mentally retarded adults.

Bertha E. Lawrence
Berkshire Community College, MA

Like many mothers, I began to look at my life after my children graduated from high school and to wonder what I could do to fill the void they had left. I was involved in volunteer work at a hospital, where I was director of the volunteer program. I was able to transfer this experience to paid employment in the social services field, but I soon realized my need for additional training.

When you've been out of school for more than 20 years, the idea of returning to the classroom can be frightening. The thought of being in a classroom with students younger than my own children left me in a state of panic. But I knew that if I wanted to continue in my chosen field, I would either have to overcome my fears and enroll in college, or be content to remain in an entry-level position.

I met with the director of admissions at Berkshire Community College; with his help and encouragement I enrolled in the human services program. It proved to be a most interesting two years. The professors were very supportive and encouraging, and the students seemed most eager to share their interests. They made me aware that I possessed knowledge that was useful to them.

After graduation from BCC, I was encouraged to enroll in the Antioch master's degree program for nontraditional students. After I received this degree I began a job search. Again my community college was there for me: the college personnel director arranged several job interviews for me, which provided opportunities for role playing and constructive criticism. These experiences gave me the confidence I needed for the job interview for the position I have today, as Home Care Program manager at Elder Services of Berkshire County. Another goal was achieved thanks to the care and concern of the members of the college community.

My thanks to the many role models and friends at Berkshire Community College who saw my potential and who extended themselves beyond the call of duty to give me the support and confidence to achieve my goals.

> *Bertha E. Lawrence graduated from Berkshire Community College in 1980 with a degree in human services. A resident of Pittsfield, MA, she manages a home care program that serves the elderly, juvenile offenders, and high school drop-outs. As a result of her concerned counseling, many young people have completed high school and gone on to lead productive lives.*

Dan Minahan
Tarrant County Junior College, TX

When I started at Tarrant County Junior College in the fall of 1973, I had decided on a career in journalism. I wanted to get my feet wet, so to speak, in a smaller college close to home and learn the ins and outs of life on a college campus. TCJC gave me a chance to do that. It also gave me a chance to see that my first choice of journalism was not what I wanted to do with my life.

I therefore changed my major to interpreting for the deaf, as I myself have a physical disability. I felt that because of this disability I could understand in a small way some of the problems the deaf population has.

As I took classes toward my goal, I found that I wanted more than to just interpret for the deaf. I wanted to help them in all skills they and other disabled people needed to know to live. That led me to change my major once more to liberal arts. In that way, I could go to a four-year college and obtain my bachelor's degree in rehabilitation. Only in getting this degree could I help my fellow disabled persons as I wished to do.

If it had not been for my years at Tarrant County Junior College, I might not have gone into a field that has given me so much enjoyment. There was no other place, at that time, around the area that offered a degree in sign language—so I might never have found my place in life if it had not been for Tarrant County Junior College.

Dan Minahan graduated from Tarrant County Junior College in 1981 with a degree in education. He now lives in Arlington, TX, and works as a vocational evaluator for the handicapped. His involvement with and contributions to handicapped individuals are reflected by numerous honors, including the University of North Texas Alumni Spirit Award, the C-CAD volunteer service award, and the George Hurt Scholarship from the Dallas Spina Bifida Association.

Alexandra Molina
Norwalk Community College, CT

I arrived in the United States in March 1978 to bring my terminally ill mother back to our native land, Colombia. What I did not know then was that I was to become a new immigrant first, and later a proud citizen.

My mother's medical condition did not allow her to go back home, and I was to remain in the United States to take care of her. My hopes and dreams and my life would have to be put on hold for a while. Difficult months awaited me in a new land. I had to cope with a new language, new culture, and a terrible illness that consumed my mother slowly and painfully.

I had left behind my husband and son, family and friends, and school. My new concern was to work, and so I got a job. My family soon joined me in the United States. One day at work I came across a brochure from Norwalk Community College. As I looked it over, a feeling of hope and adventure came over me, although the thought of going to school at that point seemed almost insane. What if I tried, and what if they accepted me? I could...maybe. I thought about it day and night, and one day I called. They scheduled an appointment with a guidance counselor, my first of many.

"Your English needs polishing," he said, and given the circumstances at home, he suggested that I should take only one course. He encouraged me to start, and so I did in the spring of 1979 with an English course. The support and guidance I received day after day allowed me to begin laying the foundation of my career.

The faculty of Norwalk Community College guided me with a patient yet firm hand. They inspired me to study hard so I could be proud of my own work and want and need more education. They taught me to fly when I was ready; they gave me my wings with the notion that the sky is the limit in the pursuit of my dreams.

With the encouragement of my adviser, I continued my education, received my baccalaureate in human services, and began to assist immigrants in family reunification and resettlement. After a few years, in 1988, I became a student once again, and once again it was with the encouragement of the faculty at NCC. This time I was a proud graduate student enrolled in a master's of social work program.

Today, I continue to receive advice and encouragement from the faculty at Norwalk Community College. They did not open doors for me; they taught me how to open them for myself. For that I will be eternally grateful.

Thank you, Norwalk Community College. I could have not done it without you.

> *Alexandra Molina graduated from Norwalk Community College in 1983 with a degree in human services. Her home is in Stanford, CT. She serves as a family services case worker within the Connecticut Judicial Department's Family Division. As president of the Stanford Hispanic Task Force and member of several community task forces and boards, she has contributed substantially to the welfare of her community.*

Mary M. Moran
Metropolitan Technical Community College, NE

Today, as a nontraditional graduate of Metropolitan Technical Community College, I reflect on my journey toward self-awareness, and how I was enhanced because of the variety of golden opportunities available to me through my educational experience at Metropolitan. Fulfilling the business management program requirements allowed me to do three things to be successful in life: to face some losses; to gain desired new information; and to take some risks. My successful educational experience was a journey of self-revelation, self-affirmation, and self-actualization.

I enrolled in Metro with a bachelor's degree and master's degree already under my belt. Unfortunately, I had never learned mathematics beyond an eighth-grade level nor been exposed to English composition. Humiliated to the core, I mourned so large a hole in my so-called education; two-thirds of the three R's were missing from my repertoire—a devastating self-realization!

Although initially humiliated, I began with characteristic grim determination my journey into the mysterious world of numbers at the developmental level. Simultaneously in English Composition I, I learned the theory and practical process of writing. I sweated through the summer session, punching the daylights out of my calculator and creating crumpled paper mountains. However, the semesters passed and so did I, garnering four "A's" in six math courses and producing "A" work in three writing courses. The challenging process increased my confidence and affirmed my belief in the God-given ability to learn and to use information.

Risking to succeed, I used all of my wonderful new information to venture into competitive academic events. Though not chosen, I first applied for recognition as Phi Theta Kappa Honor Student of the Year. Then I wrote up (complete with five-year financial projections), submitted, and presented my business plan for statewide competition; the award was fourth place. Finally, carefully crafting my words to the 450-word limit, I applied for a national fellowship award. Those 450 words won the award, and I anticipated, enjoyed, and survived a rigorous 10-week internship this summer in Washington, D.C.

These unanticipated self-actualizing events were the fruit, the evidence of successful living: of progress along the journey of self-awareness. Today, as I look forward to an uncertain future, I also look backward to draw from my educational experience at Metropolitan Technical Community College. I am strengthened knowing I can face some losses, I can gain some new information, and I can take some risks. Experience assures me that I have the tools for successful living, and for that I am grateful.

Mary M. Moran graduated from Metropolitan Technical Community College in 1989 with a degree in business management. A resident of Omaha, NE, she serves as youth program director and executive director of the Action Committee for Social Change. Among her community service activities, she has worked with battered women and given speeches, shelter, and transportation to these women.

Jose Muñiz
San Diego Mesa College, CA

As a young man growing up in San Diego during the 1960s, I thought of higher education as a thing for the rich. I never knew what higher education meant, much less what college involved. Like so many of my friends during my senior year in high school, I was confused, scared, and felt empty about my life.

I had no direction or goals. All I knew was that I would soon be faced with a big decision—what to do with my life. Much of this fear and confusion was due to lack of support from counselors and teachers to continue my education beyond high school.

During the 1960s, people of color had few opportunities. It was a period of civil unrest and the Vietnam War. During this time, I spent three years in the army as a helicopter crew chief, and nearly two years of this was in Vietnam. Luckily, my early survival training, growing up in the barrio, aided me in getting through that time of my life.

A year after my army discharge, and at the urging of my mother, I enrolled in Mesa College, located across town in a neighborhood very foreign to me. I was surprised that support, the key ingredient so lacking in my formative years, became evident my first day on campus. Information poured out of students and professors alike. From the first day, I was hooked on education and my new-found school and community.

My two years at Mesa College proved to be the groundwork for my years spent in service to the community. The intellectual stimulation I received forced me to challenge myself for what lay ahead in human care services.

Upon my graduation in 1972 with an AA in Chicano studies, I went to a state university, prepared to tackle its challenges with a deeper understanding of the meaning of higher education. For the first time, my goals were clear and I would breathe a little easier knowing that something positive was happening in my life.

For me, education and community service go hand in hand; there is no separation. Armed with an AA, a BA, and 15 years of community service, I promised myself to never allow what happened to me to happen to others. Service and more service is the answer to breaking down the barriers of ignorance and poverty. Fortunately, I'm able to do this in my career, as well as in volunteer efforts. In my position with the GAIN program, I'm able to assist welfare recipients in getting job training and in finding employment. The countless volunteer opportunities have allowed me to serve thousands of students and community members in an even greater way.

I thank Mesa College for enabling me to help others in my personal and professional life.

> *Jose Muñiz graduated from San Diego Mesa College in 1972 with a degree in Chicano studies. He now lives in San Diego, where he is the employment program representative to the GAIN program. This program assists welfare recipients with job training and employment counseling. The community service organizations to which he has devoted volunteer time and leadership number more than a dozen.*

Michael P. Murphy
Berkshire Community College, MA

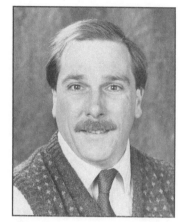

Schoolwork was never hard for me. What was hard was not knowing what came after school.

After high school, I worked as a laborer. However, my fiancee, Sue, was insistent that it was time for college. With very limited finances and little knowledge about enrolling in college, I decided to try Berkshire Community College.

While it was slow going at first, all I had to do was take advantage of what the community college offered. I always enjoyed working with people, so I chose to major in human services. With the support and guidance of two faculty members in the department, I soon took off. After two years, with my grades steadily rising, I was offered a position with a day care center where I had done a student internship.

While working at the center, I completed another year at BCC during the evenings. I think the evening continuing education program cemented my positive college experience. The other students and I all had one thing in common: a desire to learn and grow.

Through the positive reinforcement given me by everyone at Berkshire Community College, I was able to see my true value. I attained my potential. I worked at the day care center for 11 years, often as the only male in the city working with preschool children. I soon joined a statewide network of men in day care and helped form a local support network for other men in the field.

In my present job as information and referral coordinator for a child care resource and referral program, I am often asked to participate on statewide committees and to help train new workers in my field. I'm well respected by other child care professionals.

When I return to Berkshire Community College and speak to students about what I do, I explain that the community college is a stepping stone to many different levels, personally and professionally. I speak highly of my days—and nights—at BCC. I have a wonderful family, a very fulfilling job, and I owe much of where I am today to Berkshire Community College. Having an associate degree has taken me surprisingly far. Some day I hope to get other degrees. BCC has shown me that, at any age, anything is possible. I am always happy for the opportunity to give a little back to the community college that gave so much to me.

Michael P. Murphy graduated from Berkshire Community College in 1977 with a degree in human services and now lives in Pittsfield, MA. He is an information and referral coordinator for a child care resource and referral program and participates on statewide committees to help train new workers in the field.

Gerald Pantaleo
Catonsville Community College, MD

My experience with Catonsville Community College began after high school graduation in 1966, when I began my freshman year, and continues today in my role as an adjunct instructor. Additionally, in my full-time position as a vocational rehabilitation counselor, I refer an increasing number of my clients to the community colleges because of the variety of programs available. I consider myself an advocate of the community college concept and would like to share the reasons for my continued commitment.

While in high school I wanted to attend college but was unaware of the options, responsibilities, and most importantly, the personal growth necessary for the successful completion of college and the attainment of career goals. The concerned staff and small classes at Catonsville provided me with the non-threatening environment I needed to successfully determine my strengths and limitations with respect to a college education. As a result of the guidance I received at Catonsville from faculty and staff, I completed my associate's, bachelor's, and master's degrees.

During the last 10 years, I have taught at Catonsville and other area community colleges. I have seen that the benefits the community college offers are truly a service to the community. Many students I have come to know at Catonsville have sought this school out to evaluate their academic abilities, just as I did, in a supportive environment. It is very satisfying to know that I have the opportunity to carry on the role of an encouraging faculty member at Catonsville.

As a vocational rehabilitation counselor, I have frequent contact with personnel at Catonsville and other community colleges, usually regarding special needs or reasonable accommodations for a handicapped student. In all instances the professionals at the community colleges treat the individual as if he or she were their only student. With this type of personal attention, motivated individuals can meet their educational objectives.

Although I have a fulfilling career in counseling and teaching, I have often wondered where I would begin to investigate additional skills for my current positions or future endeavors. Without a doubt, I would first turn to my community college colleagues because— no matter what my age or circumstance—I want to learn in a supportive, nonthreatening surrounding.

Gerald Pantaleo graduated from Catonsville Community College in 1969 with a degree in general studies. Now residing in Annapolis, MD, he works as a vocational rehabilitation specialist for the Maryland State Department of Education and as an adjunct assistant professor of psychology at Catonsville Community College. He is a strong advocate for those physically challenged by cerebral palsy, a condition with which he lives.

Linda M. Ruiz
Austin Community College, TX

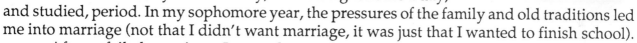

When I was growing up I recall questioning, silently, actions taken by adults (teachers, parents, prominent community persons) that hurt my childhood life. Being Hispanic, monolingual, minority, and the child of laborers, I could not always express my questions for the fear of my family; I could never doubt the childrearing method nor the discipline. That led me to grow up a loner, low in self-esteem and insecure in relationships. I also developed a deep conviction as to how I was going to do things differently when I grew up.

After high school I went to college and took 21 hours per semester at Howard Payne University, worked eight hours a day, and studied, period. In my sophomore year, the pressures of the family and old traditions led me into marriage (not that I didn't want marriage, it was just that I wanted to finish school).

After a failed marriage, I moved to Austin, TX, and began to work in a child care center. This allowed me to be a part of my children's lives. The director's job at the center allowed me flexibility. I soon realized I needed more training. In 1979 I registered in the Austin Community College child development program. That is when my commitment from childhood reappeared and conquered my life.

I first encountered the head of the department in a program management class. She was not dressed as the professors of the '60s (or as I stereotyped a professor to be); instead, she came in very casually and looked like another student in the class. She helped me overcome all of my inhibitions; I lost all my fears of being too old for school and not fitting in.

I couldn't finish in two years as I wanted because it was no longer just me, but rather me and my three little girls, debts, and a full-time job. In the first year I had to take time out for my child who had dyslexia. I felt helpless, but the support that I received from the faculty in the child care program at ACC was and has been tremendous. In the 10 years it took to obtain my degree, there was never a moment when I could not pick up the phone, call the department head, and ask for advice. If she was not there, the persons on her staff treated my concerns with respect and addressed them with professional information.

When my career was working from six to five and going to parents' meetings at the children's school, I had little time to sit and plan a course schedule. I would call the head of the department, and she would tell me which courses to take. She got me through and I finished.

Linda M. Ruiz graduated from Austin Community College in 1989 with a degree in child development. She directs the El Buen Pastor Child Development Center in her hometown of Austin, TX. Recently appointed to the City of Austin Child Care Commission, she also serves on the boards of several community agencies dedicated to providing children with positive self-esteem and a healthful environment.

All the courses opened doors for me to advocate for children's rights. I no longer am silent. I have input in the adult sector and want to make a difference for children. My commitment is to participate in the development of programs that will ensure children positive self-esteem that will nurture the body, mind, and environment. Thank goodness for community colleges that allow any person, regardless of who they are, the right to an education.

Connie Schilling
Kaskaskia College, IL

By the time I was 12 years old, I knew I wanted to be a hairdresser. So, while most of my high school friends were taking college prep classes, I was taking classes that would be beneficial toward my vocational goal.

My life was falling into place just as I had hoped. I became a hairdresser, wife, and mother, all within three years of graduation. After six years, I was no longer a wife, but a busy single parent of two pre-school-aged sons.

Even though I enjoyed my job, working in a small, southern Illinois town as a hairdresser didn't pay very much. I had often thought about taking classes to get a better job, but I always had excuses for not going ahead with it.

In March 1980, after a serious car accident, I wasn't thinking about moving up financially. My thoughts were focused on physically moving again. I had sustained a spinal cord injury, initially causing total paralysis. Despite a miraculous recovery during the next six months, I could not return to my job, so I became a client of the Department of Rehabilitation Services. The first question I was asked was, "What do you want to do now?" Well, my answer was to work, but what could I do? Since a physical job was out, I decided I had to use my head, which meant college.

Two days after being discharged, I started classes at Kaskaskia College. At my guidance counselor's suggestion, I took only two classes the first semester—human values and attitudes, and speech. Both of the instructors for those classes helped me to realize my self-worth, to regain my self-confidence, and to not be afraid to talk about my life experiences.

After that wonderful first semester, I was on my way. But not without a lot of help from caring instructors who knew that I was the nontraditional student three times over who needed a little extra guidance, encouragement, and help. When I went on to Southern Illinois University to finish my bachelor's degree in home economics education, I realized that it would have been overwhelming for me to start my education at a four-year school.

I truly feel that if I had not had the positive experience with the staff at Kaskaskia College, I would never have developed the confidence and the positive attitude I've needed to succeed in my jobs. In my present position with the Client Assistance Program, I am helping persons with disabilities to realize that they, too, can be independent and can solve whatever problems that they encounter. If only they could all have had the same wonderful experience I had in attending Kaskaskia College!

You know, a few months ago, just six years after first attending KC, I walked into the college and was greeted by my guidance counselor and an instructor, who still knew me by my first name and even gave me a hug. Now that's genuine caring, don't you think?

Connie Schilling graduated from Kaskaskia College in 1983 with a degree in education. She now lives in Carlyle, IL, where she serves as a human relations representative for Region 1 of the Client Assistant Program of the Illinois Department of Rehabilitation Services. A victim in 1980 of a disabling accident, she works now to help others with disabilities to realize that they can be independent and solve the problems they encounter.

Susan Scoda
Brookhaven College, TX

I have achieved a lot in recent years. I am proudest, however, of my Brookhaven College degree. Everyone at Brookhaven encouraged me to grow, learn, and achieve. I did not often feel that I couldn't do what I was challenged to do. In fact, I strived for excellence because the people around me believed in me. Everywhere I went, there were positive people guiding me to be the best.

Each department at Brookhaven has superstars. These were the teachers who taught more than just basics of a subject, but also helped me relate the information to my life. English with Carolyn Herron gave me a fabulous and almost gaudy vocabulary. With Steve Link to guide me, I learned how basic principles of psychology work. Joan Weston helped me apply those same ideas to understand how social structure is organized and challenged me to help change society's ills. Through the combined efforts of Johnyce Alders and Delryn Fleming, I gained the initial self-confidence to continue my education. Even the usually boring required government course was fascinating with the remarkable Phil Berry, who taught me to analyze government and look in depth at those we elect to rule.

While a student, I worked at Brookhaven as well. The Student Programs and Resources Office is staffed with three amazing women. Lou King, Carrie Schweitzer, and Joy Arndt are both friends and colleagues. Although I was just a student and a "go-fer," they welcomed me and showed me kindness and respect. When I confided to them a week or so before graduation that I had only a denim miniskirt to wear with my cap and gown, they bought me a beautiful dress. I later wore it for my wedding gown, and they were there to share in that occasion too.

I heard many excellent speakers at Brookhaven College. The Brown Bag Lunch Series gave me insights into problems I encounter every day. One speech in particular is the basis for my current life philosophy. Susan Humphries of radio station KVIL spoke of the attitude she had which allowed her to attain her present position. Her most remarkable statement was a simple one that could be used to sum up the teaching philosophy at Brookhaven: "Always say to yourself, `I can do that!' and you can achieve anything you try!"

Since graduation, I have had the courage to try several careers. I spent two years training disabled young people to live independently. I then was appointed the volunteer coordinator and tour manager for the Dallas Historical Society. All I had to say was, "I can do that!"

Susan Scoda graduated from Brookhaven College in 1987 with a degree in liberal arts. A resident of Garland, TX, she acts as volunteer coordinator and tour manager of the Dallas Historical Society. Among her community service activities have been working as an aide at the Dallas Independent School District's school for the handicapped and as a job site coordinator and instructor in that school's vocational education program.

Sara J. Snyder
Wayne County Community College, MI

In August 1974 I entered a halfway house for treatment of alcoholism. In six years since graduating from high school, I had gone from what appeared to be a normal 18-year-old to a street drunk sleeping in doorways and eating out of dumpsters.

After four months of treatment, I decided to do something with my life, so I enrolled at Wayne County Community College to earn an associate of applied science degree in substance abuse counseling.

I had been a waitress, dishwasher, bank teller, ice cream truck driver, and encyclopedia saleswoman among other things. I thought I could do anything, but I found that my alcoholism and drug addiction had harshly affected my ability to learn. I suffered from a severe short-term memory deficit and a very low tolerance for frustration.

My first year at WCCC was very difficult, but I had one instructor who told me I could do it, that I had a lot to offer. She was supportive of me throughout two and a half years of struggle and study.

People at WCCC believed in me. If I had not made the decision and commitment to go to WCCC and had returned to my old pattern of taking the first job available just to pay the rent, I'm not sure I would be alive today. What I learned and who I became because of the start I got at WCCC most probably saved my life. I don't think I would have stayed sober just "putting in my time."

In January 1986 an amazing thing happened. I walked up the front steps of the halfway house I had lived in 12 years earlier. This time, though, I had a key to the home and a business card that said I was executive director of it. Since then I have worked to make Grateful Home a caring, supportive program for recovering alcoholic women, one that believes in our strength as women and that expects us to do our best while supporting each other through periods of doubt.

I no longer look at stacks of clean dishes to prove my day's accomplishments. Today I look at the lives of women I have counseled and see their pride and joy. Two of them are now members of my staff.

I have been able to take a childhood of pain and fear and a young adulthood filled with addiction and despair, and turn them into tools to reach women who feel painfully alone. My life story can give others hope, to show them there is a way out of darkness. Without the encouraging words of the WCCC staff, I probably would have given up before my life had a chance.

Sara J. Snyder graduated from Wayne County Community College in 1971 with an AAS degree in substance abuse counseling. She now lives in Grosse Pointe Park, MI, and is executive director of Grateful Home, Inc. A recovering alcoholic herself, she now works to rehabilitate recovering alcoholic women in the same halfway house which treated her when she was an 18-year-old.

Renee Wallis
Pierce College, WA

I attended a community college in Washington when I was in my mid-20s and older than the average college student. But at Pierce College I was not alone. There were many mothers who had raised families, retired men, and people who had married and divorced. All of us wanted nothing else so much as to attend college. Though I had often felt at odds with the world, at Pierce College I found a home.

We adult learners would gather between classes in the cafeteria. We would read each other's work and discuss each other's lives. Marriage, jobs, and children mixed with Physics 114 in a mingling of the intellect and emotion, the ideal and the practical.

We went home after class to our families and the increasingly complicated situations of our lives with a growing intellectual understanding that changed everything. The unforgiving temperament of the truth often made us uncomfortable, demanding answers to difficult questions. One thing was certain; education did not make life easy.

It was a heady time. There was Magda, 45 years old and unhappily married. There she stood, clutching her English book, a German-born woman struggling through an English course that required more grammar than I, a native speaker, had ever acquired.

But we knew she would triumph. After all, my own mother, then the chair of the Social Science Division at Pierce College, had attended as an adult learner years ago. A mother with four children and no money. It was possible. It could happen, with hard work and determination. My mother was the proof.

The teachers lavished us with attention. As adult learners, we were suspended in a special state of animation. We were our teachers' adoring disciples, and the teachers were our examples of all that was possible. They knew that we could accomplish much, and their expectations were high.

Because of the years I spent at Pierce College, I know there is a place in the world for difference. I learned to keep as many doors open as possible, because who can say how someone is most comfortable entering a room? I work now as a fund-raiser, raising money to provide opportunity for others. Just as I traveled a circuitous route, just as Magda needed her chance to push through, I work to keep the door ajar for others. And I strive to follow the example set by my experience at Pierce College. Work hard. Aim high. Dream big.

Renee Wallis is a 1985 graduate of Pierce College who now lives in Arlington, VA. With the skills she acquired in her arts and sciences studies, she now works as a fund developer and has successfully helped numerous nonprofit organizations dedicated to assisting the less fortunate.

Mabel Word
Southside Virginia Community College, VA

I would never have imagined that I would be asked to write what I have been through in the last five years. After 12 years of marriage, when I was totally dependent on my husband for all means of support, he left. He said he was going out of town to traffic court and would return on Friday. He never came back. I was left with three daughters, a 10th-grade education, and no financial security. I searched all over town for employment with no results. I needed help. The only place left was the Department of Social Services. I remember the feeling I got every time I went there. I felt a little less of myself with each visit. However, through the Social Services Department, I was referred to a GED preparation program sponsored by Southside Virginia Community College. I started attending classes and I began to feel better about myself. Even at times when I didn't do well, I was motivated to study harder.

I wish I could say that going on to higher education was an easy decision, especially after completing the GED preparation class and passing my GED test. I entered SVCC with the uncertainty of whether I was going to make it. I had to learn to drive and get my driver's license. Until I got my license, I hitchhiked the 21 miles to campus and 21 miles back on the days I couldn't find a ride. The majority of students were much younger than I. However, the resources available at SVCC made my transition much smoother. I began to realize that whether we were young or old, we were all there for the same purpose, to acquire knowledge and independence.

I wish that others who have been through what I have been through would think of their children, if not themselves, where education is concerned. I feel that our offspring tend to follow in our footsteps, and our failures become their failures. By our accomplishments, we can set an example for them. If we believe in ourselves, there is nothing we can't do.

I received my associate degree in human services from SVCC in May 1989. I am currently employed at the Sheltered Workshop in Farmville, VA, a rehabilitation service agency. I instruct mentally and physically handicapped clients in vocational skills. The job is both fulfilling and rewarding. I hope that by sharing my experiences, I will be able to help someone in a similar situation.

Mabel Word graduated in 1989 from Southside Virginia Community College with a degree in human services and now lives in Farmville, VA. She is an instructor at a sheltered workshop for retarded adults, where she instructs mentally and physically handicapped clients in vocational skills.

CHAPTER EIGHT

Business

Paul Abbott
Lehigh County Community College, PA

There are many popular misconceptions about community colleges. Some will say that the community college is merely an extension of high school. That is not true. Others feel that there is no real college atmosphere because most students commute. That is not true. Some claim the academics are not as rigorous as at four-year institutions. That is not true. As a former student of both Lehigh County Community College and a four-year college, I feel I can fairly evaluate my community college experience. I received a thorough, personalized, challenging education. That is true.

I will tell you what else is true. Let's talk quality, affordable education. Four months after I was discharged from the armed forces, I knew it was time to increase my knowledge and prepare myself for a career. I sought guidance from the community college and was surprised at the enthusiastic assistance from the director of admissions, academic counselors, professors, and the financial aid staff. They went well beyond the usual assistance to help me make the decisions that were right for me.

There was no such thing as an easy course at LCCC. I worked very hard to earn my degree there, and I was well prepared to continue my studies at a four-year institution. My courses at the community college transferred with no difficulty to my four-year school, because I had planned the transfer in advance.

My two years at LCCC were two years well-spent. I developed intellectually, socially, and personally. Campus involvement for me has extended well beyond my two years at school. For the past 12 years I have been actively involved in encouraging people to attend the community college. The alumni association also keeps me in touch with those who have graduated.

The community college made an important, positive impact on my life. My education has allowed me to positively impact others' lives. As far as I'm concerned, that's what education is all about.

Paul Abbott graduated from Lehigh County Community College in 1977 with an associate degree in criminal justice administration. He now lives in Allentown, PA, and is safety coordinator at Stanley-Vidmar. While a student, he assisted in organizing "Unity," the first Black student organization on campus, and served as its first president. Since his graduation in 1977, he has served as president of the Allentown branch of the NAACP and represented LCCC as minority and nontraditional student outreach representative in the community.

Marie F. Alexander
Rappahannock Community College, VA

As a child I always thought I would go to college. One of my childhood dreams was to attend the well-respected, historical College of William and Mary, and I spent my high school years preparing for it. My dream was waylaid. As many young people do, I put aside one important aspect of my life for another: marriage and a family. During the next 15 years I was employed in various vocations: wife, mother, secretary, dental assistant, and manager.

My personal life had been through many drastic changes, and a time of uncertainty and confusion settled on my home. A time for decisions had come. After much deliberate thought, I decided to rejoin the working market. After several months of applications, I realized in order for me to re-enter my business field, I needed to learn new skills and techniques and renew old abilities. After hoarding special funds through the summer, I enrolled in Rappahannock Community College in the fall of 1986.

The greatest challenge was walking through those doors. I was 31, a mother of three, an estranged wife, and scared.

The next three years held accomplishments, pitfalls, and developments. Upon learning of a gender equity program, the Equal Opportunities for Employability Program, I was able to stay in school and complete my associate's in applied science degree in office systems technology. At a time when I needed assurance the most, I was given the opportunity to prove my ability and self-worth.

Each completed assignment led to a completed course. Each completed course was one step closer to my chosen degree. I know what it is to be elated. I know what it is to be disgusted. I especially know how it feels to be totally drained, too spread out, wanting to walk away. This always happens near the end of a long-term goal.

I will always be indebted to Rappahannock Community College and the instructors who were patient and willing to exert a tireless effort with me. Their work and mine has paid off in many ventures, including my present position at Dominion Bank. I expect it to continue to bring in dividends, possibly to include attending the College of William and Mary.

I have not lost my excitement or my thirst for knowledge. What a wonderful thing to learn when I thought all my decisions were made years ago.

> *Marie F. Alexander graduated from Rappahannock Community College in 1989 with a degree in office systems technology. She now lives in White Marsh, VA, and is an executive secretary for the Dominion Bank of Richmond. She was a displaced homemaker who took advantage of a program at the community college that helped her overcome hurdles and realize a childhood goal.*

Bessie Anderson
Los Angeles Southwest College, CA

As one of a family of 12 children from the South, and with my mother a midwife and father a farmer, it was a struggle to accomplish <u>any</u> goal. But I have always had a dream to go to college, be successful, and keep the family name strong.

I attended a junior college in the South and obtained an AA in business administration. I moved to California in 1962 in hopes of continuing my education and obtaining a master's degree. After attempting to enroll in a university in California as a junior and being told that the college I had attended was not accredited, my spirits were dampened, and I decided to just stay away from college. But, as I observed others taking classes and advancing all around me, I was motivated to pick up the pieces and continue my education.

Eight years passed. In 1971 I enrolled in Los Angeles Southwest College. It was a real struggle, working and being a housewife and a mother. There were times I could only enroll in two or three units per semester, but with the support and encouragement of the instructors, family, and friends I received an AA in 1979. Because I was still working, a housewife, and now a mother of two, the motivation to continue was even harder to muster.

I decided to go into the banking industry and progressed all the way to operations manager. In 1980 I enrolled again at LASC and took courses related to skills management and technical learning, which helped me get a vocational teaching credential for employment with the Los Angeles Unified School District. Several years later (1986), I left the banking industry, re-enrolled in LASC, and continued to pursue my goal in early childhood education. I found that LASC had an excellent program. The efficient instructors, along with the availability of the college in my immediate community, made it all worthwhile and helped me obtain my second AA in June 1988.

I now have a family day care center (Bessie's Christian Inspired Day Care), and I have continued to pursue my goals in Gospel song writing, singing, and acting. I am presently an actress in the well-known drama "Precious Lord." I have traveled extensively as a singer, and I am known as the "Little Lady with the Big Voice."

I owe so much to the instructors for their push for excellence at LASC. They helped me to attain many goals that were almost lost. I love you, my alma mater.

Bessie Anderson graduated from Los Angeles Southwest College in 1988 with a degree in child development. She now resides in Los Angeles, where she owns and operates a day care center. In addition to offering quality child care to meet the needs of citizens in her community, Ms. Anderson sings, writes songs, and acts in numerous artistic productions throughout Southern California. Her positive outlook and performing style have uplifted the lives of many people.

Michael C. Austin
Mohawk Valley Community College, NY

As valuable as my experience was as a student at Mohawk Valley Community College, I learned some equally valuable things on the other side of the desk when I had an opportunity to be an instructor in the college's evening division a few years ago.

You've worked all day, and now you're going to be out once a week until 9:30 or 10 o'clock at night to teach. And of course you need to prepare for class, at least another evening of your time. It's hard work if you're of a serious mind about your job. Lesson one.

After you've done it for a few semesters and you file your tax return at the end of the year, you find the tax return has shrunk by close to the extra amount of income you made teaching. I'm not sure if that's good or bad, but it's lesson number two.

Lessons one and two add up to lesson three: You're doing it because you really enjoy it. I always thought highly of my MVCC instructors, even more so after going on to a four-year college. But now I had a chance to look at them from another perspective. I think it gave me new appreciation for the community college concept and the value of what it does.

Add to that the most pleasant part of the community college experience, and it seems to hold especially true in the evening division: not only are you there teaching because it's something you want to do, but the students are there to learn because it's something they want to do. They may be working a job just to attend college, or they may be working a job they don't like and are going to college to start a new career. People playing on their own coin are highly motivated. They learn a lot. And there is no substitute for a community college's ability to bring in teachers who can give students the value of their practical experience to their students, not just book theory.

And speaking of practical experience, I'll relate the advice my father once gave me. When I was 16 years old and entering college, I thought I was the smart one, and here he was handing me another dad-ism. I wanted to go to a big-name school, but my parents, with six kids to raise, hadn't money available for big-name school tuition. "You're going to get out what you put into it, Michael. Go to MVCC. It's a fine school," my father said.

I got a lot out of MVCC. It's a fine school. Smart man, my father.

Michael C. Austin graduated from Mohawk Valley Community College in 1974 with a degree in liberal arts and sciences. He now lives in Utica, NY, and is director of corporate communications for Utica National Insurance Group. He oversees all company publications, is involved with corporate advertising and public relations, and handles special projects, including production of a company film for the firm's 75th anniversary. His work has won many professional awards.

Elington I. Bates
Northern Virginia Community College, VA

I have never met a broom that I liked. Immediately after high school, I dove head-first into the job market. My experiences varied from disappointing to degrading. I responded to numerous help-wanted advertisements only to be asked that dreaded question, "What can you do?" Invariably, the broom was discussed. It seemed that before I was to go anywhere, I had to "sweep America first."

Reluctantly, I accepted a job with broom at a local Chevrolet dealership. Admittedly, I didn't pursue the job with a great deal of enthusiasm. I can recall a service writer remarking that I didn't sweep with ambition. Little did he know, my ambition was not to sweep at all.

The summer of 1967 marked my first year since high school graduation and the expiration of my tolerance for the broom. I joined the Army. Four years later I was back into the job market, but neither I nor the market had changed. My immediate response was to find a quick fix. I enrolled in the Automotive Diagnosis & Tune-up Certificate Program at Northern Virginia Community College. I intended to take a few automotive courses, get a certificate, and be on my way. Fortunately, my quest for the quick fix was countered by a steadfast commitment to education in the person of Herbert E. McCartney, an associate professor in the NOVA engineering technology division. Following Mr. McCartney's advice, I enrolled in the associate degree program. Thanks to the faculty and staff at NOVA, instead of a quick fix, I joined the NOVA family at the ninth annual commencement with an associate's degree in automotive technology.

NOVA changed my life. The college switched my thinking processes from stand-by to on, and my expectations rose, both of which served to obliterate my self-imposed limits.

NOVA as an intervening factor presented both short- and long-term effects. The long-term effects bond the student to the institution and to the pursuit of higher education. I have maintained a course of personal development, which includes a bachelor of arts degree, participation in community activities, and providing a positive effect on those I touch. I measure that effect by the thank-you card from a seventh-grade class I recently spoke to, or the plaque that simply states, "Thanks Coach Bates," or the "thank you a thousand times" letter from the dean of student development for participating in the spring education tour, or better yet, a letter from NOVA President Richard J. Ernst, stating, "Thank you for serving as our commencement speaker...."

The community college experience, dollar for dollar, is richer and more rewarding in terms of breaking the cycle of poverty, despair, and disadvantage than any other program ever funded.

Elington I. Bates graduated from Northern Virginia Community College in 1975 with a degree in automotive technology. He now lives in Herndon, VA, and works as a field manager in the Audit Department for Chevrolet Division of General Motors. He is a certified youth sports coach, member of Herndon Optimists, and NVCC's member of Who's Who in American Junior Colleges. He serves as a part-time automotive lecturer at NVCC and was keynote speaker at the college's commencement ceremonies in 1988.

Merle R. Blair
Independence Community College, KS

The words will always be with me. Words that move me as much today as they did in 1955 while participating in the Independence Community College production of Romberg's <u>The Student Prince.</u>

Golden Days in the sunshine of a happy youth, Golden Days, full of innocence and full of truth; In our hearts we remember them, all else above; How we laughed with a gaiety that had no sting—we will know, life has nothing sweeter than its springtime; Golden Days, Days of youth and love.

Later I would find life does have sting, and later I would fully appreciate how Independence Community College had prepared me to deal with life's sting and still accomplish goals.

My father died in the summer before my senior year in high school. Going to college was something I always knew I would do. I had learned in the Independence school system that you can't accomplish much without the help of a lot of people who care about you. Looking back, I now realize I enrolled at ICC because I knew the instructors and coaches were people who cared about individuals.

It's not possible to name them, but as a group, the ICC administrators, faculty, and coaches were great role models. As ICC students, we learned we are responsible for the quality of our lives and the shaping of our destinies. Abe Lincoln said, "I'll prepare myself, for some day my opportunity will come." Only in retrospect can I appreciate how well ICC prepared me for the many opportunities that were ahead.

Athletics has been an important part of my life—not so much the games, but the lessons of sportsmanship, teamwork, competition, fairness, and many special memories of coaches, players, victories, and losses—all ingredients of the game of life. My ICC coaches were and still are my inspiration. I knew they cared, and I didn't, and still don't, want to let them down.

We are the sum of our life's experiences. ICC gave me the opportunity to experience the excitement and rewards of learning; the thrill of athletics, playing in an orchestra, and acting on the stage; and my first introduction to broadcasting in a radio class. So, because of ICC it's;no coincidence I went on to graduate from Washburn University in Topeka, played football and basketball, had lead roles in two Shakespeare stage productions, and later became chairman of the Washburn Board of Regents. And the ICC introduction to radio class led me to become general manager of two Topeka radio stations just eight years after ICC graduation.

I am forever grateful for those Golden Days at ICC, but more important, grateful for preparing me for the Golden Days of family, career, and life after Independence Community College.

Merle R. Blair earned an associate degree from Independence Community College in 1955. He now lives in Topeka, KS, and is the president and CEO of the Greater Topeka Chamber of Commerce. While in college, Merle belonged to Phi Delta Theta and served as a member of the Sagamore Honorary Men's Leadership Society. He has also served as a member of the Washington University Board of Regents, the board of directors of the Jayhawk Council of Boy Scouts, and the board of directors of the Topeka Public Schools Foundation.

Cora B. Brown
Jefferson Community College, KY

The fond memories I have of my two years at Jefferson Community College will be cherished forever. What I best remember is the caring, nurturing atmosphere in which we all were encouraged to excel. These memories have enabled me to meet the challenges of life as well as the setbacks, for they dwell deep within and provide comfort when it is most needed.

Since my graduation from Jefferson Community College in December 1977 and from Bellarmine College in May 1979, the value of the lessons learned has become increasingly important. The execution of my responsibilities as an assistant vice president for Liberty National Bank often requires knowledge that can only be obtained through a study of the liberal arts. Fortunately, I received an excellent education in the liberal arts while at JCC. As a working mother, I have found that knowing a little about a lot can be quite useful. Moreover, the skills valued most in my service to the community are usually those that require a well-rounded education, as opposed to those skills for which technical proficiency is mandatory.

Ultimately, the specific ways in which my life has been enriched are not only difficult to quantify, but are difficult to articulate as well. However, one particular incident remains vivid. During my second year, I received a full-tuition scholarship from the Greater Louisville Claims Association. When the elation wore off, I realized I was petrified at the thought of having to enter a room full of strangers to accept my award. Little did I know that Dr. Ainsworth, who I suspect sensed my discomfort, had decided to attend. It is often said, "It's the little things that mean so much." My memories are filled with many of these precious moments. Among them are my nomination and induction into *Who's Who Among American Junior College Students*, as well as my participation in the Health Education and Welfare Grant which studied junior college students. I am a better person because of these experiences—fond memories indeed.

Cora B. Brown graduated from Jefferson Community College in 1977 with a degree in arts and now lives in Louisville, KY. She is assistant vice president of Liberty National Bank. Cora received a Black Achiever Award in 1982, was nominated for Outstanding Young Woman of the Year in 1985, served as vice board chairperson for the Community Health Center of Western Louisville, and has held officer positions in the Kentucky Black Women's Forum and Kentuckiana Women's Network.

Lynne B. Byrd
DeKalb College, GA

Back to school at 42! Like most of my generation of women, I had married and started my family at an early age. Opportunities for women were not as open to us as they are today, or at least we didn't imagine them to be. I was a good student in high school and had one year of college before marriage. In the back of my mind I always hoped that one day I would be able to go back and get my degree.

As the last child left for college, I decided to give it a try at DeKalb College, a community college in my area that is an arm of the Georgia State University System. Here I was with two grandchildren, and I was just as terrified of registering as I had been at age 18. The college had a special program for returning oldies called Second Wind. I went to an orientation program and heard success stories from others who had returned to school.

Thus began an exciting four years for me. It was possible for me to work full time and attend night classes, so it took this long to complete my two-year degree. The small campus and its accessibility and affordability were perfect for my situation. I relearned good study habits, found excellent counsel for my planned degree, and thrilled myself with good grades. I enjoyed brainstorming and exchanging ideas with the younger students who had just entered from high school. I made some lasting friends among these acquaintances. I developed new eyes for the world in geology, new ears to hear French, an expanded understanding in philosophy, and I reveled in the English and literature courses I had so looked forward to taking. My husband was my biggest cheerleader, helping me study and celebrating my good grades. I became active in Second Wind myself, encouraging other older students to return to school. At graduation I received my associate degree in general education with highest honors.

My experience at DeKalb has made me an insatiable learner; my taste for education has just begun. I am now enrolled at Georgia State slowly working my way toward a degree in English, my first love. In my case, my desired degree will be a means to an end. I enjoy a very successful career in real estate, but my learning experience is feeding my soul and realizing a dream I have had for many years. DeKalb College helped me to realize this dream, and I am working with the alumni now to help promote the opportunities that DeKalb has to offer. My advice to older students is that it is never too late to learn. Don't delay—go for it!

Lynne B. Byrd graduated from DeKalb College-North Campus in 1987 with an AS degree in general education. She now lives in Georgia and is a successful real estate agent in the Atlanta area. She is especially active in soliciting support from the local business community for scholarships to deserving students and also in making these scholarships known to potential students.

Chris Callis
Harford Community College, MD

When my second daughter was almost two, I decided to enroll in an evening accounting class at Harford Community College. When asked what program, I said business administration; it seemed appropriate, since I had been doing bookkeeping for 15 years and needed a challenge. I had no specific goal at first; I just wanted to test the water.

Well, it was great. The professors were interesting and supportive. I continued with night classes while working and raising my two daughters with the support of a wonderful husband. It took seven years to get my AA in business administration, but it was worth it. At graduation, to my surprise, I received the *Wall Street Journal* Award and Harford's Student Achievement Award.

At this point I almost stopped, but an accounting professor (the first teacher I had at Harford) called to ask me if I was transferring to a four-year college. I said after seven years I thought I'd give myself and our finances a break, but he advised me that with my 4.0 average, a scholarship would be available if I would go full time. I decided to give it a try, and two years later I had a BS in accounting.

After taking the review course, I passed the CPA exam. This prompted me to enter public accounting with an international accounting firm, where I am now an audit manager.

Looking back at my first years at Harford, the positive experiences encouraged me to continue, formulate goals, and achieve a professional position that was beyond my highest expectations. I have always been involved in community service, and I have had the opportunity to serve on the board of the Harford Chamber of Commerce as director, secretary, and treasurer. I was appointed by the county executive to serve a three-year term for the Private Industry Council. As a charter member of our Lioness Club, I have served as director, secretary, vice president, and president. On a professional level, I am immediate past president of the Baltimore-Washington area chapter of the American Women's Society of CPAs.

In addition to giving me opportunities to serve my community, my experience at Harford set an example for my daughters and made them aware of the importance of education and community involvement. My oldest has a bachelor's degree in business administration, and my 19-year-old is a junior majoring in graphic design with a minor in sociology.

Without the ease of transition from wife-mother-bookkeeper to college student that Harford provided, none of this would have been possible. Thank you, Harford, for opening these horizons for me.

Chris Callis graduated from Harford Community College in 1976 with a degree in business administration. She now lives in Fallston, MD, and is an audit manager (CPA) with an international accounting firm. Her scholastic and professional accomplishments are complemented by her extensive involvement in the Harford County community. Her willingness to volunteer her services and expertise to the local Chamber of Commerce, Private Industry Council, and Lioness Club has improved the quality of life in her hometown.

Richard H. Campbell
Eastern Maine Technical College, ME

As I reflect on my academic experiences, I would never call myself a scholar. Even today I consider my best assets to be common sense and an ability to work with my hands. Interestingly enough, that's how my public school teachers had classified me and, in fact, guidance counselors later advised me not to take the SAT, saying that I was not college material.

Fortunately, I overcame early weaknesses in math and science, and a newly formed two-year college, offering programs of study in occupational education, gave me a chance. What an enlightening experience! For the first time I found education to be worthwhile and actually fun. After a stint with the Air National Guard, I returned, more mature and better focused, to Eastern Maine Technical College. My general education courses, especially, and the personal attention of my instructors helped me come to terms with my life and, for the first time, with the issue of aspirations.

Now, with the technicalities of my education under control, I became involved in the student community and its politics. I ran a successful campaign for the presidency of the student body, was elected to the student senate, and served as an elected representative to the statewide Technical College Student Committee.

Certainly the growth I experienced and the encouragement I received from the faculty and staff at Eastern Maine Tech contributed as much to the development of my professional life as the technical education I earned.   In 1974, having started my own one-person drafting company, I immersed myself in solar energy. By the late 1970s, having independently studied energy-efficient, solar-tempered concepts, and with the benefit of national HUD solar design and construction grants, my company became a recognized leader in residential construction. By the early 1980s, in response to local challenges and economic pressures, I had expanded into commercial and office construction, and today circumstances have extended me into development of residential housing projects, hotels, and recreational condominiums. At the same time the R.H. Campbell family of companies has grown by three divisions: millwork, interiors, and insulation.

After 16 years of modest successes, I readily acknowledge the effect Eastern Maine Technical College has produced on my aspirations, my abilities, and my career.

Richard H. Campbell graduated from Eastern Maine Technical College in 1970 with an AAS degree. He is president of his own general contracting firm. In addition to his accomplishments as a leader in the building construction field, he is past president of the Bangor/Brewer YMCA Board of Directors, a member of the boards of trustees of Bangor Theological Seminary and John Baptist Memorial High School, and an incorporator of the Eastern Maine Medical Center and the Bangor Savings Bank.

Mary Smedley Collier
Valencia Community College, FL

Why in the world would you want to go back to school?" How many times my friends asked that question when, just a few years ago, I announced my intentions to attend college. It was a legitimate question. I was in my mid-50s, had four grown children, and worked full-time in the family business. Yet, even though my days were full, I found my life devoid of any feeling of real personal accomplishment. A college education was the one thing I had always yearned for, and I knew that the time had come. Without having an answer to everyone's questions, I set off on what would prove to be one of my life's great experiences. It was an experience in learning, in growing, and in being.

From the outset, I sensed something special about Valencia Community College. That special "something" was the instructors. They made me feel my education was of the utmost importance to them. At every turn, I was encouraged to pursue my goals, offered help to achieve them, and praised for my efforts. After completing the first session, there was no holding me back. I began to move forward. Always, the instructors were excellent. And, oh, how I learned. I loved it!

As the days became weeks and then lengthened into months, the thrill of anticipation never left me. With the beginning of each session, I found in it a new challenge and a new discovery. I discovered the world of humanities, the intrigue of the earth's sciences, and the depth of self-knowledge as expressed in creative writing. The changes were subtle, but they were there. My world and I were growing. I loved it!

The changes were not only in growing but in my overall attitude toward myself. Through the encouragement of my instructors, I began to accomplish work of the highest calibre. Many honors came my way. However, more important than the honors was the self-esteem that I began to feel. I felt a greater sense of value to myself, to my family, and to my community. I had become the person I always hoped to be: educated, aware, and self-confident. I loved it!

The years at Valencia were wonderful. My life expanded in so many ways. I now possess a broader view of life, and I am proud of what I have accomplished. I could never have attained my goals had I not been a part of the Valencia experience with its high standards and its dedicated instructors. Valencia will always be my academic hometown. I grew up there, and I loved every minute of it.

Mary Smedley Collier graduated from Valencia Community College in 1982 with an associate's degree in English. She now lives in Orlando, FL, where she is the comptroller of Collier Jaguar and also the owner of a pro shop. She was instrumental in the adoption of a new policy that allowed transfer students from community colleges to be considered for enrollment at four-year universities on the same basis as students already enrolled at the university. She has served on the Sports Committee of the Greater Orlando Chamber of Commerce and has been a member of the Valencia Alumni Association and the Valencia Foundation.

Raymond Donnelly
Suffolk Community College, NY

It is often said that "the journey of 1,000 miles begins with the first step." This phrase takes on special meaning when applied to the first steps taken by the many thousands of successful graduates of our nation's community and junior colleges.

My own experience was at Suffolk Community College from 1972 through 1974. Like many high school graduates of that post-Vietnam pre-Watergate era, I was looking for a starting point. Uncertain of how to find my way, I enrolled at the local community college.

The two years I spent in pursuit of my associate degree were among the most enriching life experiences I have had. Educationally, socially, and athletically, I was able to develop a strong sense of self-worth and maturity. This development continues, of course, and is still aided in strong measure by the foundation I was able to put in place with my community college experience.

What was the difference and why was I able to benefit so much from my time at Suffolk Community College? The answer is simple—the faculty. At the community/junior college level the faculty do what they are paid to do, which is teach. This is an important distinction when you are talking about the development of the individual—the primary role of the educational experience.

In recent years it has been my privilege to serve the college in a number of volunteer roles on alumni councils, development committees, and so forth. This has given me the opportunity to observe first-hand the ever-expanding role of the community college in serving returning students, minorities, and two-year career programs.

It is from the perspective of both student and citizen that I feel strongly about the contributions made by community/junior colleges to the success of individuals and the national educational system.

> Raymond Donnelly graduated from Suffolk Community College in 1974 with a degree in general studies. He now lives in Coran, NY, and is a partner in The Daretel Group. In the past he has worked at RCA, GTE, LM Ericsson Telcom, Rockwell International, and Telenova, Inc., a subsidiary of Wang Laboratories. His loyalty to SCCC is evident in his work on the Alumni Council. He has served as its former secretary and is currently its vice-president.

Jim Douglas
South Seattle Community College, WA

When I went back to school at South Seattle Community College, I had no master plan for success. In fact, it had been a long time since I had experienced success in school. My educational achievements were more of an evolutionary process.

Learning was fun when I was a young child, but I lost the excitement of learning and a lot of self-confidence when my parents divorced during my grade school years. Junior high school was a complete write-off, and high school was not much better. I now joke with Jerry Brockey, the president of South Seattle Community College, about the days when he was my high school health teacher and would kick me out of class because I was so disruptive.

After high school, I joined the military service. When I got out, I had a series of blue collar jobs. I was working for Sears in 1974, selling automotive products, when the oil crisis hit and my income dropped to half of what it had been. Still struggling with self-esteem troubles, I seemed to hit bottom in my personal life as well at that time.

In the depths of that crisis, I wasn't sure where I was headed, but I knew I wanted to change. I lost weight, changed jobs to become a night janitor, and moved in with my mother. I visited a counselor at SSCC who helped me start in the college transfer program.

Because of a scheduling mix-up, I missed the first day of music class, and my old bad feelings about school returned. I explained to the music teacher that I didn't think I would be a very good student, based on my past experiences. She told me she graded on attendance, and if I would just show up, I would pass the class. I learned to love music. I joined the SSCC Community Choir and participated in performances. I joined my church choir and I'm still singing.

The counseling and that first class at SSCC helped me recapture my love of learning. It was at SSCC that I got the idea that education is a lifelong process. This has continued in my adult professional life. After SSCC, I went on to earn my bachelor's degree in business administration at Seattle University in 1976 and my MBA there in 1981. Not a day goes by that I don't use some part of my education, both the learning skills and the growth I experienced at South Seattle.

I have tried to pass on support to others who may have had a lapse in their formal education. A community college instructor believed in me when I was starting out and that made a huge difference in my life. I want to pass that on to others.

Jim Douglas graduated from South Seattle Community College in 1975. He now lives in Seattle WA, where he is an assistant vice president at Continental Savings Bank. Jim is active in civic concerns and believes in challenging others to achieve their potential. He is a member of the boards of the SSCC Foundation, Camp Colman, and the Southwest YMCA District Facilities Committee. He is also active in the Seattle University Alumni Association and is organizing an SSCC Alumni Association.

Nancy L. Duda
Tompkins-Cortland Community College, NY

I was the oldest of seven children growing up in the 1950s and 1960s, and little emphasis was placed on the importance of education in our home. It took both parents working full-time to keep their brood clothed and fed. An honor student who dreamed of college, I was encouraged by many of my teachers, but was regretfully informed by my parents that I could not expect their help to further my education. When I graduated in 1966, college was a far-away dream, obtainable only to rich kids whose parents could afford to send them.

Soon after high school I was married. The next 19 years of my life were spent raising four children and trying to cope with a marriage that was not made in heaven. At age 36, my marriage was over, and I was faced with the frightening reality of taking care of myself and my children with nothing more than a high school education—no job skills and no self-esteem.

When I entered Tompkins-Cortland Community College, my self-esteem was very low. A bad marriage can have a tremendously damaging effect on a person's emotional and psychological well-being. But I worked hard—and with every "A" I received on a test, with every teacher's smile of approval when I answered a question correctly—I could feel my self-confidence rise.

I was living off welfare, and my schooling was completely funded. To me this meant doing my best not to fail a society that was supporting my second chance. The PACE Program (Public Assistance Comprehensive Education) at Tompkins-Cortland gave me guidance, financial help, and counseling to help me cope with everyday problems. The COOP Program helped me find employment on campus. The instructional labs helped me develop strong skills so I could understand math and accounting and do well in both fields.

In May 1988 I graduated with a GPA of 3.93 and associate degrees in business administration and in liberal arts-math/science. Four months later I was chosen the Outstanding Adult Student of New York State. I was proud of my accomplishments. College was hard, and every new obstacle brought with it a fresh supply of fear and failure. But the most fantastic feeling comes when you keep pushing, defeat the fear, and end up doing more than you ever thought possible.

Today I have a career in telemarketing sales. I deal with people all day long and love it. Five years ago I could never have handled such challenging work. It was Tompkins-Cortland—the caring teachers and staff, and the financial support I received to keep me going—that has made this all possible. I am no longer afraid to want something. I know now that if you want something, are willing to find the route to travel, and work hard for it, anything can be yours.

> Nancy L. Duda graduated from Tompkins-Cortland Community College in 1988 with dual degrees in business administration and liberal arts. She now lives in Marathon, NY, and works as a telemarketing sales representative. Four months after graduating from TCCC with a GPA of 3.93, she was named the Outstanding Adult Student of New York State.

Dixie L. Eng
Northern Virginia Community College, VA

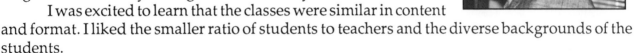

There was never a question of whether or not I would go to college. I had always been a good student and college was the next logical step. Originally, I did what most of my friends did. I applied to several universities and was accepted by all of them.

I decided to go to college away from home and chose a business curriculum. In my first year I realized I did not enjoy college life and decided to move back to my hometown. Since I moved during the school term, I chose to enroll in Northern Virginia Community College to continue my studies.

I was excited to learn that the classes were similar in content and format. I liked the smaller ratio of students to teachers and the diverse backgrounds of the students.

I was able to combine a full-time job with my education. I felt that practical work experience was important, as well as necessary, to finance my continued education, and NVCC encourages and assists students in career placement.

In two and a half years, I completed two associate degree programs: applied science in business management and applied science in merchandising-fashion.

To continue into a four-year program, I transferred my credits to The American University in Washington, DC.

My studies at NVCC and my work experience helped prepare me for working full-time while attending my last two years of concentrated study at AU. Two years later, I graduated with my bachelor of science degree in business administration-marketing.

While the community college is actually quite large and offers so many choices to students, I found the programs challenging and professors concerned about their students successfully meeting their educational goals.

I am always happy to share my experience at NVCC with others, since I feel NVCC is a wonderful beginning or bridge to a new career or outlook. I continue supporting NVCC as an advisory board member of the Business College Marketing Program. In June 1987 I was selected as one of three alumni to deliver the commencement address.

Dixie L. Eng graduated from Northern Virginia Community College in 1980 with dual degrees in business administration and merchandising/fashion. Now a resident of Springfield, VA, she is an image consultant and the manager of the Wyndham Bristol Hotel in Washington, DC. She continues supporting NVCC as an advisory board member of the Business College Marketing Program. In June, 1987, she was selected as one of three alumni to deliver the commencement address.

As rooms division manager at the Wyndham Bristol Hotel in Washington, DC, and as manager of my own business, teaching skin care with Mary Kay Cosmetics, I have found that education and learning are ongoing. When I reflect back on my structured training in the classroom and combine the knowledge with practical application, I feel the choices I made in my education have allowed me to be happy with my career choices.

Gloria Villasana Fuerniss
Ohlone College, CA

Ohlone College had just opened its doors in 1966-67. Classes had not yet started, so they didn't have any students. College officials needed some part-time student workers, so they went to a local high school, Newark High. My counselor asked me if I would like to apply for a part-time job after school and for the summer. I was hired immediately after the interview and found the work fun.

When classes for the college were going to begin, only students who went to Ohlone could work part-time as student help. In order to continue to work at Ohlone, I registered for classes. Prior to my working at Ohlone, I had not thought about continuing my education. I don't recall what I was planning to do when I graduated from high school.

At Ohlone I was a full-time student, involved in student government, business manager for the college newspaper, statistician for the football team, and part-time worker in the public information office of the college. In high school I had not been involved in extracurricular activities.

My experience and education at Ohlone helped me grow and become an active participant in my community. My parents encouraged my growth, but faculty, staff, and administrators at Ohlone gave me the confidence to succeed.

I am currently vice president/manager at Fremont Bank as well as trustee for the Fremont-Newark Community College District (Ohlone!).

> *Gloria Villasana Fuerniss graduated from Ohlone College in 1969 with an associate's degree in secretarial science. She now lives in Fremont, CA, and is vice president of the Fremont Bank. She has served as an elected trustee of the Fremont-Newark Community College District since 1979. She has served the community as chair of the City Civil Service Commission, director of the Boy's Club, adviser to the Tri-Cities' Children's Center, chair of a Boy Scout fund drive, and director of the Hispanic Chamber of Commerce.*

Marta M. Googins
Aquinas Junior College at Milton, MA

After graduating from Fontbonne Academy in 1961, I chose to continue my education at Aquinas Junior College at Milton, MA. Upon my graduation from Aquinas, I felt I was prepared to assist in our family-owned catering firm. My responsibilities required good organizational skills, the ability to professionally coordinate all types of occasions, hiring and managing approximately 140 staff persons, ordering food, and authorizing payroll procedures. During this time, I also became a registered bridal consultant and a justice of the peace— the start of my renaissance period.

For the next 22 years, I worked and became a part of a company that is one of the largest off-site catering firms in our area. I married and raised four children, who are now 24, 21, 18, and 13 years old. During those childrearing years, although I had the inner desire to return to school, it was not my first priority. Instead, I dedicated myself to the activities of my children and to involvement in my community. I was instrumental in forming the successful booster club for our high school, which was dedicated to all students in academics and athletics. Even with all these activities, I still felt that someday I would return to school and complete a bachelor's degree. I had always felt that Aquinas instilled in me the strong motivation needed to fulfill my goal.

After more than two decades of hoping to return to school, I entered U-Mass, Boston in 1985, taking a full course schedule including summers. I continued working full-time at my job, but my responsibilities at home were lightened with the help of my husband and children, who enthusiastically supported my return to school. This year, I am a senior and intend to graduate this coming summer with a bachelor's degree in sociology and a minor in law and justice. As part of my internship for the pre-law program, I volunteered 15 hours a week this past summer working with battered women and children and doing legal advocacy work within the court system on their behalf. At this time, I am volunteering 10 hours a week for the entire school year at the local police station, learning hands-on the procedure of the police system.

For the past 25 years, I have felt proud to identify myself as an Aquinas graduate, and because of that special foundation, I hope to enter law school shortly after graduation from U-Mass this summer.

> *Marta M. Googins graduated from Aquinas Junior College at Milton in 1964 with an associate's degree from its executive secretarial program. She now lives in Braintree, MA, where she operates her own catering business. She is a true renaissance woman: owner of a business, mother of four, completing a bachelor's degree, and planning to begin law school. Additionally, she volunteers in a program for battered women.*

H. Clayton Hamilton
Miami-Dade Community College, FL

I came to Miami-Dade Community College, South Campus, in the mid-'70s, upon completion of active duty in the U.S. Navy. To put it mildly, I was the epitome of the underachiever in the Navy.

Miami-Dade Community College taught me the meaning of the words commitment and involvement. The college took a relatively bright, completely unmotivated, slightly angry young man and offered him the opportunity to channel those energies into productive, worthwhile activities.

While attending M-DCC, South Campus, I was elected vice president of the Student Government Association and president of a number of student organizations. As a result of the parliamentary skills and the sense of confidence that I developed at M-DCC, I have been able to carry that sense of involvement and commitment into the Miami community at large.

As regional manager for the past five years of a subsidiary of Travelers Insurance Company, I have been able to give something back to the staff and faculty of M-DCC by offering 4036 tax-sheltered annuities to assist them in reaching their retirement goals. I also serve as vice president of the Miami-Dade Alumni Association.

Upon leaving Miami-Dade Community College, I went to Florida International University, where I majored in finance/international business and was elected president of the student government association. I recently served as vice-chair of an FIU task force whose goal was to raise $100,000 in scholarship monies for Black students. At last count we had raised $145,000.

Last year as vice president of the Dade County Sickle Cell Foundation, I chaired our annual walk-a-thon, which raised over $33,000 to assist the University of Miami Sickle Cell Center to provide screening, education, research, and counseling services to people affected by sickle cell disease.

I was recently appointed chairperson of the United Way's Core C Agency Review Panel. This panel, in conjunction with the Greater Miami Jewish Federation, provides funding to social service agencies serving the Miami Jewish Community.

Whatever success I have been able to achieve in life, whatever little bit I have been able to give back to my community, would never have happened had I not been blessed with the opportunity to begin my educational career at Miami-Dade Community College. I will be eternally grateful.

H. Clayton Hamilton graduated from Miami-Dade Community College in 1977 with a degree in political science. He now lives in Miami, FL, where he is the regional manager for Copeland Companies, a subsidiary of Travelers Insurance Company. He has served for two terms as vice president of the M-DCC Alumni Association. He recently served as vice chair of a Florida International University task force that raised $145,000 for Black student scholarships. He has also served as vice president of the Dade County Sickle Cell Foundation.

Lance H. Herndon
LaGuardia Community College, NY

Twelve years ago, I was one of several young, aspiring graduates of LaGuardia Community College who were ready to tackle the challenges of corporate America. Over the course of my professional career, I have obtained certain successes and reached specific goals that can be directly attributed to my LaGuardia experience.

Since 1980 I have served as president and owner of ACCESS, Inc., a data processing firm that provides information systems consulting throughout the Southeast. The firm grossed well over $2.4 million in 1988. The initial energy required to begin this venture came from the encouragement and inspiration that I received from both the faculty and staff at LaGuardia. I received particularly strong support from Professor Herman Washington, who took a personal interest in my progress and provided much useful advice. My undergraduate experience gave me a sense of self-worth and confidence in my abilities that enabled me to capitalize on an entrepreneurial opportunity.

My LaGuardia experiences were not only helpful in establishing ACCESS, Inc., but they also played an integral role in my receiving my first job. The services of the Secretarial Science Department provided the avenue through which I mailed resumes to all of the Fortune 200 Companies in New York City, where I received the initial training that helped me launch my career.

Finally, the Co-op Program offered at LaGuardia was also instrumental in shaping my attitude toward business. Through the Co-op Program, I learned that there is a give-and-take system that exists in our society, and as we in corporate America take from the mainstream, we have the obligation to give back to them in order to maintain a mutually beneficial relationship. As an entrepreneur, I have committed myself to give back through my service to the community. I have truly prospered both personally and professionally through my volunteer commitment to various local non-profit organizations in Atlanta, which further exemplifies the effect that the experiences of attending LaGuardia Community College have had upon my career.

Lance H. Herndon graduated from LaGuardia Community College in 1975 with a degree in data processing. He now lives in Atlanta, GA, and is the president and CEO of ACCESS, Inc., an information systems company that he founded. It is the largest Black-owned data processing consulting firm in the Southeast, with $2.4 million in sales last year. He is deeply committed to giving back to society through participation in many service organizations, including the YMCA, Operation PUSH, and the United Negro College Fund.

Kenneth Hill
Trident Technical College, SC

Throughout my years in high school, I planned to attend a four-year university and get a degree in business management/administration. After graduation, I would either work for my father in his business or enter the corporate job market.

My father bought his business in 1968 when I was 11 years old. After school, I would walk to his business, where I would play and do small jobs, such as sweeping the floor and emptying trash cans. As I reached my teenage years, I began to participate in sporting activities and lost interest in the business. But I always had a summer job waiting for me—unlike most of my friends. Now, as I look back on these years, I really appreciate my father giving me these opportunities.

When I was a rising senior in high school in 1975, I became a lot more involved in the business. My father was building the business and did not have anyone working for him whom he could rely on. I decided not to go off to college, but I did realize the importance of continuing my education.

I enrolled at Trident Technical College in the fall of 1976. I worked during the day at my father's business and attended classes at night. At times, it was very difficult (especially while taking accounting courses). However, I also think working and attending school at the same time was to my advantage. Many times I was able to apply what I learned in class at night (especially management classes) the very next morning in our business. This also raised my interest in Trident. My grades were much better than when I was in high school.

Another thing I enjoyed about taking night classes was the fact that many of the instructors worked for some of the large corporations or government agencies in our area. They had experiences they shared with us in class that were very interesting and educational. One economics instructor (who last year became the youngest senior vice president ever at his large corporation) had a great impact on some of the successes I've had in life. He constantly offered me encouragement and told me I had a very good future in the business world. He also encouraged me to continue my education, which I still plan to do some day.

The experiences and education that I received while attending Trident will never be forgotten, and I will always be grateful to the instructors and staff there.

> *Kenneth Hill graduated from Trident Technical College in 1979 with a degree in management. He now lives in Moncks Corner, SC, and is vice president of Hill Tire & Auto. He currently serves as a member of the Trident Technical College Alumni Board of Trustees and as chairman of the Board of Deacons at Macedonia Christian Church.*

Joan Jenstead
Waukesha County Technical College, WI

Waukesha County Technical College is a major priority in my life for a lot of good reasons. When I took my first accounting class, I certainly never expected to graduate with an associate degree, take many more continuing professional courses, and someday become chairperson of the WCTC District Board of Directors. But I did.

That first class provided the motivation I needed at the time. As a working mother of three, I decided to improve my skills in the accounting field in order to open up my career possibilities. I went on to complete the associate degree, since I recognized both the personal and professional value of it.

Going back to college was initially difficult, but, with the help and support of WCTC instructors and my family, I was able to juggle my college, work, and home schedules. My instructors, especially, took the time to work around any of my special situations. The environment at WCTC, a two-year college, was much less threatening than I had anticipated. The instructors and staff have a genuine concern about helping individual students develop their potential and providing the support needed by so many adults returning to college.

My accomplishments in life have been quite fulfilling, not the least of which was my graduation from WCTC in 1973. Throughout my professional life I have continued to take self-improvement courses. WCTC laid the foundation for that and helped me appreciate a meaningful education. In addition to the quality education I obtained, the college helped me attain the levels of self-confidence and self-esteem necessary to succeed in the workplace.

I progressed up my own career ladder, culminating as the president of National Realty Management in Brookfield, WI. That initial accounting class started something big!

In 1989, I redirected my career and have my own part-time consulting and personal investment business. I decided it was time to become more involved in community activities in addition to my career.

I am quite active in a professional real estate organization, but WCTC again is a priority in my life. I have been on the WCTC District Board of Directors for over five years and served two of those as chairperson. Recently I became vice president of the Wisconsin Vocational, Technical, and Adult Education State Boards Association and serve on its legislative committee.

It feels good to be able to give something valuable back to a place that's been so important in my life—Waukesha County Technical College.

Joan Jenstead graduated from Waukesha County Technical College in 1975 with an associate's degree in accounting. She now lives in Brookfield, WI, where she operates her own private consulting and personal investment business. She has been a member of the WCTC Board of Directors for five years and currently serves as vice president of the Wisconsin Vocational, Technical, and Adult Education Board's Executive Committee. She has received special recognition as a Woman of Distinction in Waukesha County and, in 1988, was recipient of the WCTC Alumnus of the Year Award.

Stuart Klovstad
Fergus Falls Community College, MN

The major reason I attended a community college was cost. If the community college system did not exist, I would have attended college with more economic hardship, and I am convinced I would have received a lesser education.

I was very pleasantly surprised at the quality of the instructors and education at the community college. At the time, I was very naive and thought the big universities had the best teachers. Only later in life did I realize the big universities sometimes only had the best researchers. Teachers at the community college are there to teach. They brought me into their hearts and sometimes into their homes. When I left the community college, I was convinced I left with a good education at a bargain price. This was confirmed at the school I transferred to; the counselor stated that I had one of the healthiest transcripts he had ever seen. I would like to thank all of my teachers for giving me a strong academic background.

The community college also taught me how to treat individuals. As part of the work-study program, I joined the janitorial staff and cleaned the college center. In working with the janitors, I found out they treated everyone from the students, administrators, and teachers to the president with the highest degree of courtesy and respect. Likewise, we were treated with a great deal of respect. I learned a very important lesson: no matter what titles or position a person earns in life, a person is still a person and should be treated no differently from anyone else.

Community college also allowed me to compete in collegiate sports. I learned how to play tennis at the age of 16, and two years later I was playing at the college level. The important lesson I learned was that not everyone can make it to Wimbledon, but there is plenty of opportunity to find your niche in life and to be successful at a lower level.

In short, the community college gave me a strong academic background, taught me how to get along with other people, and gave me an opportunity to be a collegiate athlete. To my school, Fergus Falls Community College, thank you, for you were the very best.

Stuart Klovstad graduated from Fergus Falls Community College in 1978 with an associate's degree in business. He now lives in Fergus Falls, MN, and is supervisor of the auditing department at Otter Tail Power Company. In addition to being active in his church, he contributes to the community through Jaycees, the Big Brothers program, coaching basketball in the city's program, and serving as a member and treasurer for the Fergus Area College Foundation.

Juanita Gillard Murrell
Williamsburg Technical College, SC

In the summer of 1984 it became apparent to me that I was not going anywhere in my job. Because of my lack of education and training, I had no chance of advancement. I decided that I was going to have to go back to school to realize any progress in my life. Since I was interested in business, I looked for a college in my area that offered an associate degree in business.

One of my friends told me of the business program at Williamsburg Technical College. I visited the college and was favorably impressed. The staff at the college was friendly and answered all of my questions. I found I could get the training I needed there, taking most of my courses at night.

I enrolled at the community college and have never regretted it. The instructors were excellent, the classes interesting and provocative. I was delighted with how much I was learning. My instructors at Williamsburg Technical College motivated me to the point that I decided that after I got my associate degree, I would then pursue a four-year degree. And that is exactly what I did.

In January 1988 I received my bachelor's degree from Limestone College. Deciding I needed even more skills, I proceeded to go on with my education, and in January 1989 I earned a master's degree in business administration from Central Michigan University. Now I am employed as a vendor scheduler/purchasing agent. My goal, however, is to pursue a Ph.D. and become a certified purchasing agent.

Currently, I also teach part-time at the community college in my area. I feel I have much to teach those who are planning to enter the business world. I use myself as an example to those beginning their education. All journeys begin with just one step, I tell them. My journey to my goals began with Williamsburg Technical College. I will forever be grateful for the tremendous boost it gave me on my long climb. Now as I look back I can truly say, "Thank you, Williamsburg Technical College. You taught me that no dream is too large, no hope can't be fulfilled."

Juanita Gillard Murrell graduated from Williamsburg Technical College in 1986 with a degree in business. She now lives in St. Stephens, SC, and works as a purchasing agent. Ms. Murrel is secretary of her PTA and vice president of her church. She is also involved with the Boy Scouts of America and is a member of a sorority that does extensive social work.

James J. Palmeri
Northampton County Area Community College, PA

Personal Service. Those are words that are very important to me in business. They are also words that I recall when thinking about my experiences at Northampton County Area Community College.

I enrolled in Northampton's new Funeral Service Program in 1977. Only 21 students were enrolled in this specialized area of study. We had moved into a newly completed building designed specifically for the program. We all had a close bond. Other students in the school looked at us with intrigue, as they had many questions concerning the funeral industry. We became almost a novelty. Yet with the novelty came interaction, because everyone wanted to be part of the funeral service class. We were close-knit, and people wanted to be a part of us.

Since there had not been a funeral service school in the eastern part of Pennsylvania for many years, many potential funeral directors were waiting for the chance to go to school. The average age for the class was the mid-30s. There were two wives of funeral directors, two daughters, and a number of sons hoping to get a license to expand their families' businesses. There were also a number of us "independents" who were hoping to start our own funeral home some day. (Fortunately, I was one of those who was able to realize that dream in a short amount of time, opening my funeral home a year after graduation in my hometown of Martins Creek, PA.)

The lessons that I learned at Northampton were practical and personal. No one can "teach" you to care for someone; that's a natural thing. But, when everyone around you cares and is willing to go an extra mile to do the right thing, the attitude wears off on you. The instructors at Northampton care. Funeral service was a new thing. Curricula had to be developed to mold this new type of student. General study instructors had to learn what our profession was all about. The two years went by quickly. The instructors and the students had an unbelievable camaraderie in searching within each other to see where the other was coming from. It was fun, but it was work.

Personally, I wanted to give something back to the school that gave me the opportunity to become an independent businessperson in my community. That is why I agreed to serve on the Northampton County Area Community College Alumni Council. Later, I was elected president of that organization. That is when I truly met the "personal service" people who cared enough for the college to come back and help in any way they could to keep this college as successful as it has been and promulgate that good reputation to the future. If it was not for Northampton, I would not have been given the opportunity to meet such people and help keep the attitude that has made all my ventures successful.

James J. Palmeri graduated from Northampton County Area Community College in 1979 with a degree in funeral service. He now lives in Martins Creek, PA. He operates four businesses owned under the family corporation and, with his wife, also owns a travel agency and the James J. Palmeri Funeral Home. He became involved with the NCACC Alumni Association in 1984 and has served as its president for four years.

Gracie Welsh Romo
Yavapai College, AZ

My community college experience at Yavapai College has been the most rewarding and useful of all my educational experiences.

My first experience at a higher educational institution was not enjoyable. I attended one semester at a major university right out of high school and did not want to return. I missed my hometown, friends, and lifestyle. I needed that balance in my life, so I chose to pursue other goals.

The following year, I was selected as Miss Indian America in Sheridan, WY, and traveled across the states representing the American Indians as an ambassadress.

Upon completion of my reign, I decided to pursue my educational goals. I selected Yavapai College for many reasons. First of all, the size of the college allowed one-on-one interaction with instructors. Second, a student could fulfill and transfer the basic requirements of a four-year program. Third, the college had a good Native American program, and ultimately, the total environment fit my lifestyle.

I was pleased with my selection of Yavapai College. It has fulfilled all the qualities of a learning institution that I was looking for. My instructors were wonderful people and teachers. They were available when I needed them; they were stimulating and provided encouragement. I excelled in all my courses and graduated with honors. This served as an incentive to continue my education.

The Native American program at Yavapai was beneficial to Indians as well as non-Indians. I had the opportunity to serve as a peer counselor and enjoyed helping fellow classmates. We helped students with counseling and tutoring and provided social programs. This made the transition of lifestyle and study habits easier for many students from the reservation. Other students were able to see and learn from the Native Americans through the annual Pow Wow and Pageant. I commend Yavapai College for having one of the best Native American programs in the state and would like to encourage the college to promote this remarkable program.

Yavapai provided a firm foundation of study habits and learning skills necessary for further education. The experience at Yavapai made university schooling smooth and simple. I even use those same skills and habits in my current business operation and in everyday life. I am very grateful for the experience at Yavapai. It has truly been the most rewarding and useful of all!

Gracie Welsh Romo graduated from Yavapai College in 1981 with a general education degree. She now lives in Prescott, AZ, and is owner/manager of the Yavapai Indian Smoke Shop. She was selected as Miss Indian America and traveled across the United States representing American Indians as ambassadress. She remains a leader in the local Yavapai Indian tribe and has strong beliefs about the need for tribal role models.

Nytza I. Rosado
Reading Area Community College, PA

After I graduated from high school, I wasn't sure about college. I chose Reading Area Community College because it was near my home and I could afford the low tuition.

Reading Area Community College is a small school, and the staff tried hard to help each student succeed and develop his or her potential. Students helped each other, and if anyone needed to improve study skills or was having trouble understanding a subject, tutors were available.

Part of the richness of my RACC experience was meeting students of many ages from diverse families. We learned to work together to achieve our goals. We gave each other encouragement and confidence.

Today I am working on a bachelor's degree and am deeply involved in community affairs. It was the sense of community on RACC's campus that motivated me to get involved in community affairs.

In June 1989 RACC's Alumni Council honored me for my professional accomplishments and community service by naming me the Outstanding Alumna for 1989. I was very proud to receive this award. I owe a lot to my community college and am pleased to be a graduate of RACC.

Nytza I. Rosado graduated from Reading Area Community College in 1982 with a degree in business administration. She now lives in Reading, PA, where she is assistant treasurer and branch manager for the Bank of Pennsylvania. She is a member of the Pennsylvania Commission for Women, the Board of Directors of Reading City Schools, the New Futures Task Force, and the Latin American Political Action Committee. Additionally, she is a campaign specialist for the United Way of Berks County and a youth counselor at St. Peter's Roman Catholic Church.

Steve Shearn
Fresno City College, CA

As a high school student in a small rural school, my all-consuming passion was athletics. My parents were both college graduates and emphasized education, but the focus of my life was my next football or basketball game. I held my own in the classroom but didn't really understand the importance of a solid academic background.

Like many high school athletes, I dreamed of being recruited to play college football at a major four-year school.

As I came to the mid-point of my senior year, I had to face the fact that Bear Bryant and Ara Parasegian weren't going to be calling me. Since I wanted to go to college, I now had to accept reality and select from my available options. My first option was to attend the U.S. Naval Academy, since it had recently notified me that I had been accepted. In addition, Fresno City College had a strong athletic and academic program and seemed genuinely interested in having me enroll there. I had to pick between these two schools.

The opportunity to attend a military academy is often the chance of a lifetime. I'm sure that many people questioned my decision to attend Fresno City College instead. Since making that decision 16 years ago I have had many opportunities to ask myself if it was the right move. My answer is always a resounding YES. Fresno City College provided me with the opportunity to realize every goal that I had envisioned and some that I hadn't.

The college's football program allowed me to test myself against some of the best junior college athletes in the state. I was able to surpass my initial goal of just making the team and actually excelled at my position. With this success, I finally earned my high school dream of a scholarship to a major four-year school.

My decision on a career was also made while I attended Fresno City College. I chose a major in business and found that the thrill of the classroom could be as exciting as that of the stadium. The education and experience I received in those two years became the foundation that I built my life upon.

It's been almost 15 years since I sat in a classroom at Fresno City College. I now own a pension consulting business in the city of Salinas, CA. Every day I use something I learned in my two years at FCC. Some days it's the mathematics fundamentals I was sure I'd never need. Other days the mental toughness that I learned on the football field makes the difference between success and failure. I always use the intangible people skills I picked up from my association with FCC faculty and classmates.

Steve Shearn attended Fresno City College and graduated in 1976 with a major in liberal arts. He now lives in Salinas, CA, where he owns a pension consulting business. He is an active member of his local Rotary Club and is involved in community improvement activities.

Some people may question the need for or credibility of a junior college. Unfortunately for them, they missed a fantastic experience. In retrospect, I can honestly say that making the choice to attend Fresno City College was one of the best decisions I've ever made.

Linda Shimada
West Los Angeles College, CA

I'll never forget the first day I came to West Los Angeles College to enroll. I was in my early 30s, and I was so intimidated to be on a college campus that I was actually shaking. No one could have been as scared and insecure as I was. I kept thinking, "What am I doing here?"

I never thought I'd go to college, but I knew that without an education, I'd always have limits placed on me in my career and I'd never be able to get into a management position. So I decided to take a few classes at West to give it a try, and I gathered up my courage and asked the first person I met how to get to the admissions office. I'll always remember how he showed me the way there and made me feel welcome. From that first moment, people at West were ready and willing to help me. Everyone seemed to understand how I felt, and they kept encouraging me to "go for it." When I look back on that first frightening day, I think of it as my lucky day, because my whole life seemed to get better and better after that.

I'll always be grateful to the wonderful faculty at West. My instructors stand out in my mind for being very supportive and informative. And they gave me a solid academic training as well, which is evidenced by the fact that after I completed my studies at West, I transferred to the University of Southern California, where I graduated magna cum laude and was elected to Phi Kappa Phi, the all-university honor society. It never would have happened if I hadn't gone to West, where I always got what I needed.

I've been real estate coordinator for the Culver City Redevelopment Agency for the past five years, coordinating the acquisition and relocation of properties. It's the kind of job I always dreamed about, and I definitely would not have been able to get this position without a college education. West started me on the road to a very rewarding and interesting professional life.

I also believe West helped me in my personal life. When I graduated from USC, my son graduated from high school, so I was going to school when he was, and it really helped me to identify with his problems. And I think I was a good example for him—he went on to USC, too, which he probably wouldn't have done if I hadn't had college experience myself.

Education doesn't just open you up intellectually. That's just one part of it. It opens up everything else, as well, in terms of your self-esteem and your relationships with others. A college education really is a down-payment on the future, and it's the best investment anybody could ever make.

Linda Shimada graduated from West Los Angeles College in 1980 with a degree in business and now lives in Culver City, CA. She is a real estate coordinator; president of the Culver City Sister City Association; and a member of the Asian-Pacific Women's Network, the Culver City Guidance Clinic, and the West Los Angeles College Alumni Association. She has also served as an officer in the Culver City Business and Professional Women's Organization.

Patricia Skenandore
Northeast Wisconsin Technical College, WI

I don't remember having any real direction when I graduated from high school, and following the crowd to college seemed the appropriate thing to do. I soon found out that it was work, and I guess I just wasn't ready to get really serious. After two years I quit, got a job as a teller, got married, and had a daughter. Reality set in! I began to realize that I would be working for a long, long time, and I didn't want to be low woman on the totem pole.

The only problem I had was keeping my 40-hour-a-week job and taking classes at night. I was excited to discover that the supervisory management program available through Northeast Wisconsin Technical College was designed for working students. The best part of all was that the class material was so relevant that I could apply everything I learned to my job. I found that I was able to transfer some of my University of Wisconsin-Oshkosh credits, and I doubled up on classes, which allowed me to graduate in two years instead of the normal five.

My education at NWTC allowed me to realize my potential and understand the need for a good education. Thanks to what I learned through this program and my on-the-job experience, I was able to progress into a managerial position at a major corporation. Unfortunately, as the economy would have it, the subsidiary company I worked for closed doors, but not before I was able to foresee the future and make plans for a business of my own. I purchased Norrell Temporary Employment Service in 1988. With the challenge of owning my own business, l have realized that we can never know too much and that life is a continous learning experience. Obtaining a bachelor of arts degree from Lakeland College was the next step. I was able to transfer my NWTC credits in order to achieve this goal.

I am now finding it interesting to see how our temporary employees are also learning through NWTC. They are developing the confidence and ability to become more valuable employees with the ability to improve the direction of their lives. NWTC is an essential part of making Green Bay a thriving, productive community. Thanks to my education at NWTC, I have made the transition from performing a job to achieving a career.

Patricia Skenandore graduated from Northeast Wisconsin Technical College in 1982 with an associate's degree in supervisory management. She now lives in Green Bay, WI, where she is owner and manager of the Norrell Temporary Employment Service. She uses her expertise by helping with the United Way, Wisconsin Women Entrepreneurs, and the Green Bay Area Chamber of Commerce. She is secretary of the NWTC Alumni Association and continues to encourage her employees to develop new skills at the college.

Linda Smukler Soriano
Miami-Dade Community College, FL

I began my pursuit of an associate in arts degree at Miami-Dade Community College-North Campus in 1979 with an Honors Grant. Since it was my goal to pursue a career in business, I concentrated my studies in the business curriculum.

Although difficult at times, the continuous challenge of my M-DCC-North courses kept my interest alive and led to my graduation with high honors in May 1981. I was selected as the Phi Beta Lambda Future Female Business Leader of the Year.

Based on my academic performance, and with the help of my M-DCC-North accounting professor, I was able to obtain various scholarships that permitted me to pursue a baccalaureate degree at the University of Miami. In addition, I received the Price Waterhouse Accounting Scholarship and the AICPA scholarship.

In May 1983 I graduated with highest honors. My professional career began at Spear Safer Harmon and Co. During my tenure, I had the opportunity to work on audit, review, compilation, and tax engagements.

After completing two years, I discovered that my professional interest was in the tax area. I began working for KPMG Peat Marwick, an international certified public accountant firm, in August 1985, as a tax specialist.

For the last four years KPMG Peat Marwick has given me valuable experience in a variety of industries, including transportation, insurance, commercial, and retail. I have also worked extensively with domestic, foreign, national, and expatriate individual clients. I am currently a manager within the tax department and also am actively involved as the Miami office's primary recruiter for the tax department.

I have been a certified public accountant in the state of Florida since 1984, and I am a member of the Florida and American Institutes of Certified Public Accountants. I am also a member of the Cuban-American CPA Association and am actively involved with the Greater Miami Jewish Federation.

My experience at M-DCC-North provided me with an excellent background to initiate a successful business career. This has given me the desire to contribute to the future of M-DCC and its students by serving as a member of its advisory committee for business administration and as an adjunct professor in its School of Accounting and Business. Teaching M-DCC-North students continues to be a rewarding experience for me as I work to provide students with a foundation and a direction that were invaluable to me.

I am grateful for all of the opportunities that have become available to me as a result of my association with M-DCC, and I will continue to strive for success through enrollment in an MBA program.

Linda Smukler Soriano graduated from Miami-Dade Community College-North Campus in 1979 with a degree in accounting. She now lives in Miami Beach, FL, where she is a CPA and manager of the tax department at KPMG Peat Marwick. In addition to being active in the Greater Miami Jewish Federation, she serves as a member of the M-DCC advisory committee for business administration and as an adjunct professor in its School of Accounting and Business.

Jack Spurlock
North Lake College, TX

In 1979 I received the first associate degree in real estate from North Lake College in Irving, TX. It was quite an honor for me.

When North Lake College opened its doors, it quickly established itself as one of the leading schools in the state of Texas in real estate classes. Since my primary business was real estate, I felt I needed to take advantage of the latest information in the field, and, therefore, I earned the associate degree in applied arts and sciences, majoring in real estate.

My community college experience was one of going back to school, since I had already earned a bachelor's degree from the University of Houston. While my motive for returning to college was to receive technical real estate information, the additional advantages for a returning, mature student were many—the discipline of study, the networking advantage of meeting other real estate agents, the opportunity to share life's experiences in the field of real estate and selling, and in some instances, to retrain my thinking and broaden my perspective. The benefit of the technical education of real estate (laws, financing, marketing, investment strategy, etc.) is obvious.

The confidence that I received from my degree has been overwhelming. I feel that I have benefited greatly from the education I received from North Lake College. I encourage everyone I talk to who is interested in the field of real estate to take full advantage of the education opportunities offered by this highly respected school.

Jack Spurlock graduated from North Lake College (a campus of the Dallas Community College District) in 1979 with its first associate's degree in real estate. He now lives in Irving, TX, where he is a real estate broker operating one of Irving's largest offices. In addition to serving on the North Lake Real Estate Advisory Council, he has returned to his alma mater to teach real estate courses.

Sandra G. Stephens
West Virginia State College-Community College Division, WV

WHOA...Wait a minute!! Me!? Go back to school after 18 years? I had already held several key positions. I was working at a great job and I was married and had two fantastic kids. What could college do for me? Well, grab hold of your seat and let me tell you what it has done for me.

After much prodding from my college-graduate cousin, I agreed to look into West Virginia State College and its community college component. There were so many subjects beckoning me, I took the plunge. After 20 years of work experiences, I felt I needed a degree to advance in the business world. Little did I know I was going to get so much in the bargain! I was a little apprehensive about going back to school, since I was old enough to be the other students' mother...or big sister. But I fit right in and had the time of my life.

College left me with the ability to discuss everything from Freud to computers with ease. My English classes dusted out corners of my brain I hadn't used in a while, and my love of fabrics was fed with knowledge that I use daily in my job. I just wanted to open my head and pour it all in.

I graduated from West Virginia State College with an associate degree in applied science in March 1988. But that's not all. I did it with high honors and was recognized for excellence in scholarship. I'm a member of the Founders Week Roll of Honor. During my two-year, full-time schedule I was PTA president of my children's school. Today, I speak publicly concerning career choices and education, and I am an active member of State Alumni. I love community work and was the Nitro Citizen of the Month, December 1986.

I am actively involved in my area's community awareness group. We work with the chemical industry to keep the public abreast of our daily involvement with hazardous chemicals and our local Superfund site. I'm an interior designer with a growing company, and I teach an adult education class at night. I have full intentions of returning to State for an art degree because I miss the late night study and the ambience. (And the tests? Nah!) Most of all, I want to accomplish so much more!

West Virginia State College and the community college component have given me so much! Nothing can beat the feeling when you step on stage and receive your diploma. I love the satisfaction and the step ahead that for me has turned into a fast-paced run. And I'm not done yet! Thank you West Virginia State-Community College Division and all the staff.

Sandra G. Stephens graduated from West Virginia State College-Community College Division in 1988 with a degree in fashion merchandising. She now lives in Nitro, WV, where she is an interior decorator and designer. She is actively involved in the community, serving on the Community Safety Assessment Committee, the Rock Branch PTA, her subdivision's homeowner's association, and the WVSC Alumni Association. She also serves as an adult education instructor for Putnam County.

Robert Swanagan
Lee College, TX

In 1966 I enrolled at Lee College in Baytown, TX. Unfortunately, I hadn't made the necessary commitment for organizing my time, nor had I established the study habits that are required to be a good college student. These factors, coupled with my attendance for the first time at a predominantly White college, were not a formula for success. I remember quite vividly the White woman who was responsible for discussing the results of a test that would determine possible career options, spending 15 minutes with White students and only two minutes with me. She said to me, "You could be a dancer, preacher, or maybe a construction worker." Needless to say, I only completed one semester, and then I was drafted into the Army.

In 1972 I enrolled in Lee College again. This time I was married and had matured considerably, but I was still apprehensive about being accepted as a "Black student." However, I was pleasantly surprised, as the first three men I met—Randy Busch, Emil Vogely, and Robert Spencer, who were instructors in the industrial management program—made learning an enjoyable experience. I was working rotating shift, but they made class so interesting that I was eager to go even when I was working the graveyard shift, which was 11 p.m. to 7 a.m. I was not a Black student to them, but an individual who had something to contribute. Much to my surprise, many of the other instructors displayed the same genuine interest with regard to minority students. Lee College had been revolutionized since 1966 when I first started, and so had I as a student.

In 1975 I received my associate's degree in industrial management and have since received a bachelor's degree in psychology from the University of Houston. Currently, I am employed by one of the major chemical corporations in the world and am responsible for college recruiting activities in the southwest region of the United States. I direct 12 people in the interview and selection process of engineers and chemists at the doctorate level, as well as professional and technical employees.

To Lee College and all of its faculty and staff, I say thank you for being a change agent in my life and for truly exemplifying the role of academia toward making education attainable for all.

Robert Swanagan graduated from Lee College in 1975 with an associate's degree in industrial management. Subsequently, he received a BS degree in psychology from the University of Houston. He now lives in LaPorte, TX, and is a personnel representative for Mobay Corporation, one of the world's major chemical corporations. He is responsible for college recruiting activities in the southwest region of the United States.

Homer Taylor
College of Technology and Applied Science, WV

My associate's degree from the community college component of West Virginia Institute of Technology provided the "key" to achieving my personal and professional goals. Had the community college two-year format been unavailable to me, in all probability, I would not have attempted a four-year degree and would have never advanced beyond the craftsman level.

I worked for the local Charleston newspapers for six years prior to entering the associate's degree program at West Virginia Tech, and I continued to work full-time through two degrees, advancing at work as I advanced in my education. The community college "key" took me from an hourly craft job to a staff engineering position and laid the groundwork for my future, all while I was still in school.

Working to do the best on my education, my job, and my family role all at the same time gave me an insight into my own capabilities and future that I had never envisioned. I quickly realized that with the community college "key" in my hand, I could compete on an educational and professional level that was never before possible with my high school diploma. That same "key" allowed me to recognize that with further education I could rise as far in my chosen profession as I wanted. So with the foundation of my community college degree, I pursued and earned a bachelor's degree in business from Tech and advanced to night production manager for the Charleston newspapers.

Upon receiving my bachelor's degree, I joined Scripps Howard Newspapers in Cincinnati, OH, as assistant production and engineering director, later advancing to president and general manager of Scripps Howard Supply. My last position with Scripps was as corporate vice president of supply and manufacturing.

In 1987 I joined Knight-Ridder, the second largest newspaper group in the United States, as vice president of supply, residing in Miami, FL. I actively serve several media and civic organizations such as the American Newspaper Publishers Association, Southern Newspaper Publishers Association, Miami Chamber of Commerce, Miami United Way, and the West Virginia Tech Newspaper Operations Advisory Committee.

While no one could take away my God-given talents, I am convinced that the first step into higher education was the true "key" to my future advancements and success. I tip my hat to all the community college programs in this country as real alternatives to four-year programs and institutions. I know that you serve a very real need—you certainly did for me.

Homer Taylor graduated from the College of Technology and Applied Science, the community college component of West Virginia Institute of Technology, in 1970 with an associate's degree in electrical engineering technology. He now lives in Miami, FL, and is the vice president of supply for Knight-Ridder, the second largest newspaper group in the U.S. He actively serves on several media and civic organizations, including the American Newspaper Publishers Association, Miami Chamber of Commerce, United Way, and the West Virginia Tech Newspaper Operations Advisory Committee.

Danny Uptmore
McLennan Community College, TX

Why a community college? When I was asked this question, my thoughts rambled back some 20 years when I was confronted with those life-changing issues. I am a first-generation college graduate and was not hampered in my college decision by the desires or leaning of my parents. I was reared on a small farm in central Texas during a time when technology had not reached the farm. We still chopped cotton, hauled hay, and did many other tasks that convinced me to make a career choice outside of agriculture. "So what's next?" I asked myself after my 22 classmates and I graduated from high school. As I looked for my beginning, our local community college was the best place for me.

There were several reasons I chose a community college. First, it was accessible. It would not require a relocation away from familiar surroundings. For someone who was not very sure about his place in the college experience, this factor was very important. (Since then, I have realized it is also important because it helps prevent a "brain drain" in one's community.)

Second, it was affordable. Being a person who occasionally changed his mind, I could explore many different areas of study. This aspect appealed to me because I didn't want to begin my career facing my own "national debt." And explore I did, much to the disbelief of my instructors. I may not have discovered my true calling in those years, but I did discover some areas that both the instructors and I decided were not my forte.

Third, the community college gave me the opportunity to grow. It was a good size, but it was not intimidating. I had the opportunity to become involved in many activities, to lead, to organize, and to express my views (a very important character flaw of a teenager).

Finally, I chose the community college because of the quality. The instruction was challenging. The atmosphere was pleasant. The course offerings allowed me to explore, learn, and mature.

Would I do it again? You bet! The community college experience is the best bargain in the United States. More people are taught, trained, and retrained in community colleges than in any other type of educational institution. That's why I am still involved. I want to make sure that the vision of the community college remains accessible and affordable to everyone...that we offer everyone the opportunity to grow...that we maintain the quality of the educational product. For after all, what we offer is opportunity—the opportunity to become your dream.

Danny Uptmore graduated from McLennan Community College in 1969 with an associate's degree in science. He now lives in Waco, TX, and is a marketing representative with Unisys Corporation. He has seen McLennan Community College from all angles—as a student, as president of the Student Government Association, as part-time instructor, and as a member of the college's board of trustees for 10 years.

Berta Lee White
Meridian Community College, MS

In the tenth grade, I terminated my education for marriage to a handsome young man, and I soon learned the importance of education and in giving one's best to family, society, and government.

Opportunities for learning and serving came through 4-H, followed by leadership roles in the church, civic clubs, and on boards of a local hospital and public library. I also became an executive in a family-owned telephone business.

I was elected to the Mississippi legislature, where I served 12 years (four years in the House of Representatives and eight years in the Senate). It was not until after I became a grandmother that I entered Meridian Community College.

First, I learned that MCC had a place for everyone, whatever the age, sex, color, or nationality. Everyone was special! Classmates were friends! During legislative sessions when I was late for night classes, a classmate took notes for me. Teachers helped after hours and weekends. In 1963 I received my associate's in arts degree and later my bachelor's of arts and science degrees from the University of Southern Mississippi.

As opportunities arose, I felt prepared to assume roles on the boards of the county, state, and American farm bureau federations. My confidence continued to build. In 1977 I was elected to the Board of Country Women's Council, USA and three years later to the Board of Associated Country Women of the World, an organization representing 68 countries and approximately nine million members.

I received an invitation from President Reagan to a luncheon at the White House in recognition of American Business Women in 1982 and the following year an invitation for my husband and me from President and Mrs. Reagan to a dinner at the White House honoring their international guests.

I was appointed by the Secretary of the U.S. Department of Agriculture to participate in three international trade missions: 1985 to Europe; 1986, Asia; and 1987, Hungary, Yugoslavia, and Turkey.

My interest in other nations expanded even more and I led a trade mission for AFB Women to Sweden, Austria, and Germany followed by a CWC-USA visit to Russia in 1988. This past September I served with fellow members of CWC-USA as host to 2,000 guests at the ACWW World Conference.

In addition to my family and friends, I owe much to MCC. Staff, teachers, and classmates helped prepare this high school drop-out for a variety of opportunities to serve and contribute to my place in *Who's Who of American Women* and *The World Who's Who of Women.*

Berta Lee White graduated from Meridian Community College in 1969 with an AS degree. She now lives in Meridian, MS, and recently retired as president of Hughes Telephone Company, Inc. Her contributions to her community, government, and society have come by wearing a variety of hats: civic leader, state legislator, business executive, and American representative in the Associated Country Women of the World. As a result of these activities, she has appeared in Who's Who of American Women *and* The World Who's Who of Women.

Jaleigh Jeffers White
Wabash Valley College, IL

From my freshman year in high school I knew that I wanted to become a certified public accountant. My business teacher knew that I was interested in math and brought me several articles on public accounting and the increase in women in the field.

By my senior year in high school I had decided to attend Southern Illinois University in Carbondale. However, I did not feel that I was quite ready to leave home yet. I had a good job where I worked 20-30 hours a week, and I thought I could handle it and my studies too. I decided to attend Wabash Valley College for my first two years. It was the perfect stepping stone for me.

I visited the advisers at SIU before I attended Wabash Valley, and they helped me select my first two years' courses to be sure that my transition to a four-year institution would be as smooth as possible.

All of my courses transferred, and I was even able to take many of the normal junior- and senior-level business courses at Wabash Valley where there was a small teacher-student ratio compared to the lecture-hall format at SIU.

Wabash Valley provided the perfect transition for me to get my feet wet in several areas before going to a major university. At Wabash Valley I was involved in student government and the Model United Nations and was a participant in a scholarship pageant held on campus that eventually led me to the Miss America pageant two years later as Miss Illinois.

I would encourage anyone to use junior colleges as a transition from high school to a four-year institution. They give you time to get used to the independence of a college schedule more gradually, and for many people junior colleges provide the time for making final career decisions.

A few years ago I returned to Mt. Carmel, and now Wabash Valley again plays a role in my life. I enjoy the cultural activities and sports activities the college sponsors. My husband and I usually take a class each year, from computer programming to weight lifting. For me, Wabash Valley College has become more than an educational institution I attended for two years; it is an integral part of our community and offers many opportunities we would not otherwise have available.

Jaleigh Jeffers White graduated from Wabash Valley College in 1980 with an associate's degree in accounting. She now lives in Mt. Carmel, IL, and is a certified public accountant in the tax department of Kemper CPA Group. At WVC she was involved in student government, the Model United Nations, and a scholarship pageant that eventually led her to the Miss America pageant two years later as Miss Illinois. Additionally, she is a director of the Chamber of Commerce and a member of many civic organizations.

Brenda Williams
Marshalltown Community College, IA

Why go back to college at my age? After all, our three children are at the age to go to college themselves! It was hard to picture myself sitting in a classroom surrounded by 18- and 19-year-olds, but I decided to give it a try—one semester at a time.

To my delight, and with the help of Marshalltown Community College's faculty as well as my newly acquired college friends and my family, I survived that first semester. I found that there were students my age—and even older than I—in my classes. The faculty was willing to help when I had a problem. There was a wide variety of classes to choose from. And, most of all, there were academic advisers who saw a potential in me that I myself couldn't see yet.

At the end of that first semester I had made the grade and already had set my sights on not only the Professional Office Institute (POI) certificate, but also the associate degree. I couldn't believe it was the same me who just weeks before had been so apprehensive about going back to college.

Not only does MCC offer a whole spectrum of classes in arts and sciences, career option programs, and vocational-technical programs, but the college also has many organizations that provide opportunities for student leadership, socialization, support, personal growth, and fun. In spite of my age, I jumped right into student government groups, faculty-student task force groups, the POI Club, and OWLS (a support group for Older Wiser Learners). I'm glad I took advantage of those opportunities, because they enriched my school experience, my life outside of school, and even my family life. In addition, I experienced leadership opportunities I never would have been exposed to elsewhere.

To others, taking the big plunge might not have seemed like such a giant step as it did to me. To others, MCC might seem like small potatoes because it's only a two-year college. To me, it was the first positive boost I had ever had, and even though I now have a good job, the yearning to continue my education is real—and it's all because of MCC.

Not only did this 40-plus-year-old get her associate degree with honors and her two-year clerical certificate, but she also earned the honor of being named "Outstanding Student" out of a class of 270 in May 1989. Thanks to institutions like MCC— and especially to MCC—people like me can still grow in many ways and succeed, no matter how old we are or what our goals are!

Brenda Williams graduated from Marshalltown Community College in 1989 with an AA degree and Professional Office Institute two-year certificate. She now lives in Gilman, IA, and is an office assistant and staff coordinator for Shorr Paper Products, Inc. She has served as a community volunteer and was active in student government groups, faculty-student task force groups, the POI Club and OWLS (Older Wiser Learners) while at MCC. She also was named "Outstanding Student" out of a class of 270 at her graduation in May 1989.

Wendell W. Williams
South Florida Community College, FL

I came to South Florida Community College as a 36-year-old college drop-out. I wanted to finish my bachelor's degree and to see if I could still do college-level work. I chose the community college for several reasons. I knew it had limited class sizes; it was a safe environment; it was close to home; and I could continue working full-time, which I needed to do since I was married and had a family.

After high school I attended Rollins College in Orlando on a baseball athletic scholarship. During my junior year I withdrew due to financial reasons. When I finally got back to college, family obligations made a college degree almost prohibitive, but I came to South Florida Community College, where I found all the instructors to be top-drawer. After I gained my old confidence back and received an associate's degree, I could then transfer my credits to Rollins. I finally obtained a bachelor's degree in social science.

I believe in the community college system. After high school, many students cannot afford to go to universities. With community colleges, all students have the opportunity to earn an associate degree or learn vocational skills.

Because of my commitment to community colleges, I was appointed to the Board of Trustees at South Florida Community College. I held this position for eight or nine years and resigned only to accept appointment in 1983 as a charter member of the State Board of Community Colleges. I was its third chairman and have always been first and foremost a champion of the increased funding for community colleges to meet the needs of Florida's rapid growth. This has certainly paid off, as every year the funding has significantly increased.

Away from the community college scene, I am president of the W.A. Williams Citrus Nursery, Inc. It is a family-owned and -operated business of citrus groves and citrus production. I have been chosen as outstanding member of the Florida Nursery and Growers Association; Outstanding Trustee Board Member for the Florida Association of Community Colleges; City Commissioner of Avon Park; and I have been appointed to the Highlands County Zoning Board as well as several other city and civic organizations.

I am proud to say that my three children are all community college graduates.

Wendell W. Williams graduated from South Florida Community College in 1976 with an associate's degree in business. He now lives in Avon Park, FL, where he is president of the W. A. Williams Citrus Nursery, Inc. He has served as chairman of the State Board of Community Colleges, a past board member of South Florida Community College, and member of the Avon Park City Council.

CHAPTER NINE

Public Service

Warren W. Abriel, Jr.
Schenectady County Community College, NY

I think deep down I always wanted to carry on the family tradition of becoming an Albany fire fighter. When I graduated from Albany High School in 1967, I enrolled in college. My goal was to be an architectural draftsman—something I could do on my days off from the fire department. After a miserable first semester, I decided to enlist in the navy.

After the service, I joined the Albany Fire Department. I soon realized I had to return to college and get my degree, this time in fire protection technology.

The only program close to what I was looking for was offered at Schenectady County Community College. At the time, this curriculum was a certificate in fire science. The college hoped to offer the AAS in fire protection technology in the near future. Through efforts of the students in the program, we were successful in convincing the college that the need was there.

I finished the courses in fire science with a perfect 4.0 average, but still hadn't taken the required liberal arts courses. I was gun shy from my previous college experience. With the urging of my adviser, the faculty of SCCC, and my wife, I started with the dreaded English I.

With a little hard work, attending evening classes one or two nights a week and working two jobs, I completed my degree. After eight years, I finally finished with straight "A's" and only one "B." I had reached my goal of an AAS degree and also scored well enough to be promoted to lieutenant. I'm sure the score on the lieutenant's exam had something to do with my studies at SCCC.

At that time, Empire State College in Saratoga, a nontraditional college, began a degree in fire service administration. This program allowed me the flexibility to complete my school work around my other responsibilities. Since I had had such an enjoyable experience at Schenectady and Empire State, I continued my education and earned a master's degree in public administration from Russell Sage College.

I am currently assistant chief in charge of training with the Albany Fire Department. I am also in charge of setting up our Hazardous Materials Response Team, which was the focus of my master's thesis. In addition, I serve as a tutor with Empire State College. After 15 years of study, I think I can offer the citizens of Albany something that they might not have otherwise had.

Warren W. Abriel, Jr. graduated from Schenectady County Community College in 1980 with a degree in fire protection technology, having earned a certificate five years earlier in the same field. He now makes his home in Albany, NY, and serves as that city's assistant fire chief. In addition to contributing to the public welfare through his work with the fire department, he periodically tutors students at Empire State College.

Bonnie J. Allison
Red Rocks Community College, CO

How high can a 40-year-old housewife reach? My steps started at Red Rocks Community College at a brown-bag lunch sponsored by its Women's Center. I was feeling extremely sorry for myself—my husband was successful at a new business, and the kids were leaving home and not needing me. The other women at the lunch listened to my plight, expressed some sympathy, and then said, "Get on with your life— the opportunity is here, at Red Rocks." I was counseled to seek the associate of arts degree. I did, and I also received the Colorado Scholar's Award—a repeat of an honor I'd been awarded in high school—only then I had chosen marriage instead. This was pretty heady stuff for an old lady.

And how high did I reach? I started slowly by winning an Edgewater, CO, city council election. I was still learning at Red Rocks—discipline: finishing the assignments, on time and in the correct form. I was developing confidence as my assignments received good grades; I was learning as the doors to the world's knowledge were unlocked.

How high? I ran for office again and became the first woman to complete a mayor's term in Edgewater and be re-elected two times, once unopposed.

Next, during my three terms in the Colorado House of Representatives, I had the opportunity to attend the Harvard University John F. Kennedy School of Government for State and Local Senior Executives. Me! Bonnie Allison from Edgewater. Pretty heady.

How high? I was honored as Woman of the Year in 1979 by the *Lakewood/Wheat Ridge Sentinel*, the area newspaper.

How high? I was elected state senator from my district, running unopposed in an open seat, which I now hold.

I am the most proud of two accomplishments. First, before going to the statehouse, I put into place the mechanism for the Urban Renewal Project in Edgewater. Today, 450 new jobs exist in my city, my county, and my state. Second, the Colorado Convention Center exists because I was willing to put all my energy into the legislation that will bring 2,200 jobs to Coloradoans and millions of dollars into our economy.

I remember an old lady once telling me, "Be accountable." I hope I am. And another lady said to me, "Be good to the old lady you'll be some day." Edgewater Plaza stands today, with 79 units of elderly housing.

And finally, Eleanor Roosevelt made two statements that have become my creed:

"You must do the things you think you cannot do."

"Do not ask [of] others what you are unwilling to do yourself."

> *Bonnie J. Allison graduated from Red Rocks Community College in 1977 with an AA degree in history. She now lives in Edgewater, CO, where as a three-term state legislator she has been instrumental in making possible urban renewal projects, subsidized health care for indigent patients, and the creation of the Colorado Convention Center.*

Catherine May Bedell
Yakima Valley Community College, WA

In retrospect, my whole life ties back to the years I spent at Yakima Valley Community College (then called Yakima Valley Junior College). That small two-year college was the door that opened on my future and influenced my decisions, my interests, and my progress for a lifetime.

It was 1934, and we were in the depths of the Depression; going away to college was financially impossible, and interrupting my education to look for a job that probably was not there was a heartbreaking alternative. Then the community college in my hometown opened, and so did my future.

I didn't realize it at the time, but everything that happened in my junior college years was prologue to the next 50 years. A very special teacher worked with my mother to show me how to open a magazine agency where I sold subscriptions by phone and mail, which made it possible for me to earn enough money to finish my education. Two teachers inspired me to an interest in English and speech. A particularly gifted teacher of economics and political science sparked a lifetime interest in government. From those roots of motivation, I became a teacher, a radio broadcaster, writer, and producer, and spent over 30 years in elected and appointed offices at the state and federal levels.

Thank God you were there, YVCC, to give me my first chance to a full, happy, and, I hope, useful life. I express that gratitude on my own behalf, and also on behalf of the millions of men and women throughout the United States who got their first chance at becoming successful, contributing citizens from community colleges.

Catherine May Bedell graduated from Yakima Valley Community College in 1934 with a degree in liberal arts. Now a resident of Palm Desert, CA, she has provided distinguished service to our nation as a member of Congress and as commissioner of the U.S. International Trade Commission. Since her retirement in 1981, she has been a lecturer-in-residence for colleges throughout the United States and overseas and has remained active as an international trade consultant.

Roger N. Begin
Community College of Rhode Island, RI

Like many of my friends, I was unsure of what I wanted to do with my life when I graduated from high school. While it was expected that I would go on to college, I had no definite career plans.

In the fall of 1972 I did two things that would change the course of my life: I enrolled at the Community College of Rhode Island (then known as Rhode Island Junior College), and I decided to run for elected office.

While a student at CCRI, I was elected to the Rhode Island state legislature. The first time I cast a ballot I was voting for myself!

I ended up winning that race, and for the next four years I took classes in the morning and tried to put into practice what I learned at the state house in the afternoon. It was an exciting time, and it made my learning experience at CCRI all the more meaningful.

During my two years at CCRI, I had the chance to explore a number of fields of study. Thanks to the support and guidance of the faculty and staff of that fine institution, I discovered I had a strong interest in business.

Upon graduation in 1974, I continued my education at Bryant College, where I eventually received a bachelor's degree in business administration and began a career in banking.

Much has changed since I made the decision to attend the community college. In 1984 I decided to enter government service on a full-time basis and was elected state treasurer. In 1988 I was elected lieutenant governor of Rhode Island.

In all my years in public office, the confidence and sense of direction I received during those formative years at CCRI have helped encourage me to reach out and seek new challenges. That experience is something for which I will always be grateful.

Roger N. Begin graduated from the Community College of Rhode Island in 1974. He now lives in Providence, RI, and is the lieutenant governor of his state. His agenda of issues for the people of Rhode Island in his first term includes environmental protection, services to families and children, and affordable higher education for all citizens.

Bradley Brigham
Greenfield Community College, MA

I enrolled at Greenfield Community College in a blur—in a thick fog of confusion, self-doubt, and fear. I did this immediately after high school and less than two years after becoming a paraplegic as a result of a motorcycle accident. No particular career program or area of academic discipline especially appealed to me; I knew only that higher education was beneficial in and of itself and that I neither wanted nor was able to perform manual labor. My secondary school grades and aptitude test scores were only mediocre, ostensibly due to a lack of personal initiative, proper discipline, and adequate study habits. I was supposedly bright—at least the majority of my teachers' reports over the years pointed to my potential for scholastic achievement. But without ever having applied myself in a sustained and serious fashion, the prospect of failure at GCC was a sobering thought as I embarked on my coursework in the fall of '74.

The fear of failure—of being exposed as intellectually deficient—stirred me into such a frenzy that I did little more than study during my first semester. Fortunately, this nervous diligence paid off, as my work earned high marks as well as praise and encouragement from my instructors. Needless to say, this was a euphoric time. As I became more assured of my ability, subsequent semesters at GCC found me enrolling in progressively more challenging courses. During this time I was especially flattered—and bolstered—to be asked to work in the Learning Skills Center as a tutor.

The final phase in my metamorphosis came in the form of two acceptance letters: one from Amherst College and one from Williams College—two four-year liberal arts institutions with rigorous academic admissions requirements. I decided on Amherst for my bachelor's degree and after a hiatus of four years, Williams for my master's degree.

The success of my tenure at GCC was made possible in large measure by the extreme friendliness and ready availability of the faculty and staff. Their commitment to creating a supportive and productive environment cannot be praised too highly. I think of GCC as an incubator that brought me to life and nurtured me in new and wonderful directions.

Bradley Brigham graduated from Greenfield Community College in 1976 with a degree in criminal justice. He now lives in Colrain, MA, and works for both the STAVROS Foundation and Historic Deerfield. After becoming a paraplegic during the prime of his life, he has dedicated himself to public service and now participates in numerous community organizations.

William M. Burke
Norwalk Community College, CT

Without question, the years I spent at Norwalk Community College provided many of the highlights of my college life. It was during these formative years that I received the encouragement, support, and education that strengthened my academic pursuits and motivated me to achieve.

At Norwalk Community College I was encouraged by the faculty and administration to pursue leadership positions in student activities. As president of the student government and editor-in-chief of the college paper, I had access to many leaders within the college, the city of Norwalk, and the state of Connecticut. But far more important, my experiences at Norwalk Community College gave me the encouragement and confidence to move ahead aggressively in my career.

The associate degree from Norwalk Community College helped me obtain a bachelor's degree in management from American International College and a master's degree in education from the University of Massachusetts at Amherst. With these degrees, I started serving young people in Massachusetts, and today, as founder and president of The Washington Center for Internships and Academic Seminars, I am pleased to be able to assist young people by providing programs that encourage them to strive for professional achievement and give them the confidence to respond to the challenges of our society.

The Washington Center has provided academic internships and seminar programs to over 15,000 students from over 700 colleges and universities across the country. They intern in the nation's capital in congressional, executive, and judicial offices, as well as in other exciting organizations in Washington, DC.

We provide a number of quality programs focused on the needs of our students. One of our newest programs, and one I am very proud of, is the Minority Leaders Fellowship Program. We select outstanding minority students from across the country and provide them with full fellowships to participate in an exciting combination of meaningful work experience within the challenging environment of Washington, DC and exposure to many of our nation's leaders. This program's enrollment includes many students from community colleges.

I have confidence that our society's questions will be answered resoundingly by the bright young people graduating from our education system. And there is no question, as the pages of this book show, that many leaders in our nation have excelled and reached their present levels of success due to the fine, outstanding principles that the community colleges throughout the country have inspired in them. I again applaud the work of our nation's community colleges and thank Norwalk Community College in particular for igniting that spark that enabled me both to lead and to serve others.

William M. Burke graduated from Norwalk Community College in 1967 with a degree in accounting. He now lives in Alexandria, VA, where he is president and founder of the Washington Center for Internships and Academic Seminars. This center is the largest independent nonprofit educational organization enabling students to earn college credit for internships and academic seminars in the nation's capital.

Guy Cameron
Laramie County Community College, WY

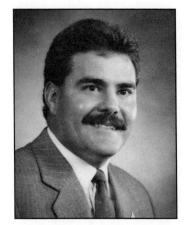

In August 1976 I started my college days at Laramie County Community College, playing basketball on a scholarship and carrying a class load of 18 hours, majoring in business administration.

I was anxious and excited to begin school. I was determined to make short work of the next two years at LCCC. Little did I know that the classes at the college level would be more intense and would take more dedication than I had put forth in previous years. Before long I found myself falling behind in classes, with basketball taking up a lot of time that should have been spent studying.

Prior to the end of the fall semester, I gave up. I just quit, determined to prove to myself and anyone else that I could be successful without a degree. I found myself unable to find a good job and ended up paying an agency to find me employment.

Four months later, while I was selling shoes, my psychology instructor, Rex Karsten, came to visit me about going back to school. What I won't forget about that visit is that Rex became more than an instructor—he became a close friend who was concerned about the choices I had made for myself and my future. But more important and memorable, Rex did not try to pressure me into doing something I didn't want to do.

I went back to school, which wasn't easy. I was going back to something I had failed once, letting down people who were counting on me, such as my parents, a basketball program, and myself.

As for Rex Karsten and Joe Phelan, a physical education instructor, their guidance did not cease with my re-entry into the community college system. I can't thank them enough for their professionalism as instructors and the individual time they each took to help me.

I graduated in 1981 with an associate degree in fire science and again in 1983 with an associate's degree in business.

I feel strongly today that had I been at a four-year institution, I would have been just a statistic. The opportunity that Laramie County Community College gave me to achieve an education right in my hometown and the individual attention and commitment of its dedicated faculty are truly the marks of a quality institution.

Representing Laramie County in the Wyoming Legislature gives me the rare opportunity to make my vote count for Wyoming's future, and I don't think that could be possible without remembering my roots.

> *Guy Cameron graduated from Laramie County Community College in 1981 with a degree in fire science—and then again in 1983 in business administration. He lives in Cheyenne, WY, where he serves as a firefighter and member of the state legislature. One of his state's youngest legislators, he has dedicated himself to enhancing the political and economic status of families through education at all levels.*

Mona Lisa Eleando
Central Arizona College, AZ

I was raised on the White Mountain Apache Nation. I am the third child in my family to go to college, but the first to graduate from a junior college with an associate's degree. Leaving the Apache Nation to attend college in the small town of Coolidge, AZ, was one of the most difficult things I have encountered. But the encouragement from my parents kept me from changing my mind.

My first day at Central Arizona College was a frightening experience. I wanted to turn around and be among all Indian students, as I was at Phoenix Indian High School. I was very shy, but I knew I had to interact with students from other cultures if I was to survive college life.

My first year at CAC was a big shock, yet challenging. Two special instructors, Dorothy Bray and Margaret Mizer, were all my reason to be there. Their words of encouragement to stay and to pursue to my highest ability made all the difference. In addition, the counseling department was there to help me when I really needed that extra boost.

As I began to adjust to campus life, things were much easier. I started to communicate with all the students—Black, White, or Mexican. I also learned to take initiative to find out what the college activities were and how to become part of them. Throughout the year, I became involved with the Indian Organization and worked as a peer advocate for the college.

When I graduated in 1982, I found that CAC had given me a desire to learn more, and I attended Mesa Community College and Arizona State University for one semester each. At that point my financial resources were exhausted. I returned to the reservation and worked as an outreach counselor, teacher's assistant, and camp director. Presently, I am the tribal vice chairman—community liaison for the tribe.

Central Arizona College had a great impact on my life. I changed from a shy person to an outgoing individual. I can communicate effectively with people on many different levels. If CAC had not prepared me, I would not be where I am today.

I am pleased to say that I was recognized as an Outstanding Young Woman of America in 1988. This is very rewarding to me, and I attribute this to CAC, as the college has encouraged and motivated me to succeed.

Believe me, college life is not easy, but giving it a try is better than not trying at all. Be a role model and be the best you can be. Good things will come out of your education if you make it a priority. Good Luck!

> *Mona Lisa Eleando graduated from Central Arizona College in 1982 with a degree in liberal arts and now lives in Whiteriver, AZ. She is a community liaison for her Apache tribe and was chosen as an Outstanding Young Woman of America in 1988. She has also contributed to her community as a teacher's assistant, secretary, camp director, and counselor.*

Roger L. Evans
Clark State Community College, OH

After graduating from the public school system, I wanted to attend college but did not do so for financial reasons. After graduation I entered the U.S. Armed Forces and served six years' active duty and two years' reserve duty from September 1954 through August 1962.

Upon leaving military service, I worked in a foundry to support my wife and two children. Remaining at the foundry only nine months, I worked for the U.S. Post Office for three months before beginning my career in law enforcement on October 21, 1963.

In the fall of 1968 I entered Clark State Community College and became one of the first of 17 students in the new law enforcement program. Going back to school at the age of 32 and working full-time as a police officer seemed at first to be an impossible task, but the enthusiastic support of my family and the college's faculty and staff contributed greatly to my graduating with honors in June 1971.

My experiences at Clark State were extremely beneficial to my career and whetted my appetite for more knowledge. My quest for excellence in my chosen profession began there. I became only the third member of the Springfield Police Department to be selected to attend and graduate from the FBI National Academy. I also attended several other schools and successfully completed various courses, including Adelphi University Community Drug Awareness Team courses, Northwestern University Police Command School, and the Federal Bureau of Narcotics and Dangerous Drug Enforcement program.

The disciplines I was exposed to at Clark State Community College prepared me to meet and deal with the many challenges public service presents. I fully believe that what I learned at Clark State helped me overcome the hurdles I have faced in being chief of police. My accomplishments in the field of law enforcement and community service have permitted me to help others become successful in their careers and lives as well.

Roger L. Evans graduated from Clark State Community College in 1971 with a degree in law enforcement and now lives in Springfield, OH. As chief of police for the Springfield Police Department, he is a community leader and role model to many people. He is a graduate of the FBI National Academy and also studied in the Adelphi University Community Drug Awareness Team, Northwestern University Police Command School, and the Federal Bureau of Narcotics and Dangerous Drug Enforcement program.

Charles E. Ford
Lansing Community College, MI

I was an All-State athlete at J.W. Sexton High School in Lansing, MI, and received the Darrel E. Rolfe Award, a special student athlete award. One of my goals was to play basketball in college, but everybody told me I was a little small. Lansing Community College Coach Art Frank came to see me, and that was my beginning at Lansing Community College, where I was the nation's leading basketball scorer, averaging 36 points per game. At LCC I studied to be a math teacher and a coach, and I will never forget Bill Petry, my math instructor at LCC. He and I hit it off, and we are real friends.

I had visions of going on to a four-year college and continuing my studies to be a math teacher. I thought much of this dream was ended when in the twelfth game of the season in my final year at LCC I broke my leg and could hardly walk for two years. There I was at 84 credits and practically suicidal. My grades had been up and I was playing well, and now I didn't know what to do.

After finally getting my cast off, I quit LCC and worked full-time for two years. But peer pressure finally began to get to me: in 1976 it dawned on me that all my friends were beginning to graduate from college and I wasn't in college anymore. Friendly competition is always good, and I began to realize that athletics are only a supplement to education, not the other way around.

I called Art Frank, and he was right there knocking on my door. I began to see that the world was still turning. I finished my six credits at LCC, and with Art's help ended up with a full basketball scholarship to Northwood Institute. I was captain of the basketball team there and graduated with a bachelor's degree in business administration.

From Northwood Institute I went on to Michigan State University, where I earned my master's degree in labor and industrial relations in December 1982. I entered Cooley Law School in 1984 and passed my bar exam in May 1989. I graduated in two years and seven months, earning the designation of Most Distinguished Graduating Senior out of 101 students. This designation is awarded for academic performance and leadership characteristics. I was president of the Black Students Association and senator of the class.

Serving an internship that gave me the opportunity to work 16 weeks with senior citizens fit right into my position with the Department of Transportation, where I have continued to work since 1972. How proud I was when I earned the American Jurisprudence for Collective Bargaining.

Serving my community and my state remains my highest priority, and I was recently elected to the Lansing City Council. I have a wonderful future with many doors of opportunity open, and this stems from the people who cared at Lansing Community College.

Charles E. Ford graduated from Lansing Community College in 1974 with a degree in general education and now lives in Lansing, MI. He is a state executive at the Michigan Department of Transportation and a member of the Lansing City Council. As an all-state basketball player at LCC, he led the nation in scoring; meanwhile, he served as both a student senator and president of the Black Students Association.

Howard Godwin
South Florida Community College, FL

Ientered South Florida Community College with the cavalier attitude of a high school sophomore, believing that I could build my race car, play semi-pro football, and continue to work in the family groves, with my college work simply falling into place. After all, that's what had happened in high school. I was shocked when I was suspended for poor grades. My plans had always included a bachelor's degree in criminology, for I had decided when I was 12 that I wanted to be a sheriff and had held to that idea. Now I was totally disappointed as well as frightened.

I immediately petitioned and was reaccepted by the school and, for the first time, I saw SFCC for what it really was, a family waiting to aid and assist me. This was an important turning point in my life. The entire staff, which was very small, helped me. They taught me how to study, new methods of comprehension, and what work responsibility meant both in and outside of class. From that time on, everything became easier for me. I received my associate degree, went on to the University of South Florida for my bachelor's, and then, my greatest joy, I went to work as a deputy sheriff. By this time, thanks to SFCC, I realized the true value of education and was able to properly align my priorities. While working as a deputy, I completed my master's degree at Rollins College.

I campaigned for sheriff and was elected this past year. I have worked harder and been happier than at any other time of my life. I hope to be a part of bringing the best sheriff's office in the state of Florida to Highlands County and to serve the people here for as long as they need me.

Howard Godwin graduated from South Florida Community College in 1976 with a degree in law enforcement and now lives in Sebring, FL. After serving as a deputy sheriff and earning a master's degree, he recently won election to the position of sheriff of Highlands County, FL. In this capacity, he has been an outstanding law enforcement officer and has been extremely active in church and civic activities.

Dwight Henry
Cleveland State Community College, TN

Simply stated, Cleveland State Community College made it possible for me to continue my education and helped me to believe in myself. Cleveland State's open-door policy and low tuition gave an ambitious but unfocused 18-year-old from the housing projects an opportunity. No one in my family had ever gone to college. Later, both my sister and mother would be beneficiaries of the same wonderful experience.

Cleveland State prepared me well for the academics of a four-year institution. Never did I feel deprived or deficient for having attended a community college. My professors were generally very caring and encouraging.

My interest in government was sparked at Cleveland State. My involvement in the student government association would plant the seed for future roles as mayor of Cookeville, TN, and as a member of the Tennessee House of Representatives. Success at Cleveland State gave me a positive self-image, enhanced hope, optimism, and self-confidence.

As the years roll by, one tends to reflect on past decisions and wonder, "What if?" It has been over 16 years since my student days at Cleveland State. Now, more than ever, I realize that attending a community college was a very good decision. Thank the Lord for those two years of growth and inspiration.

Cleveland State, I will be forever grateful.

Dwight Henry graduated from Cleveland State Community College in 1976 with a degree in history and now lives in Cookeville, TN. He is a Tennessee State Representative and president of Manna Broadcasting Co. In 1985 he was elected mayor of Cookeville and helped initiate "Cookeville Tomorrow," an award-winning effort addressing housing and road projects.

David Mathis

Mohawk Valley Community College, NY

As a teenager, I did not consider myself college material. Too often both my elementary and high school guidance counselors failed to encourage me to consider college. In my case I was not only an average student, but also a Black.

In the '60s, Blacks and other minority high school students were not encouraged to go to college unless they were exceptional in athletics. Most of the rest were pushed toward vocational training. College was something I did not really think about; no one in my family had even graduated from high school. I was the first.

After graduating from high school in 1966, I went to work as a mailroom clerk. After meeting college-educated white collar workers, I began to realize that they were not much different from myself. Then, the thought followed that if I did not want to stay in the mailroom for the rest of my life, I would have to try college. Since my employer had a program that paid for college courses, I gave it a try.

I took my first college course at Mohawk Valley Community College in the fall of 1966. My first experience at college was successful. I was convinced that I could compete, and after 10 evening courses, I took a leave from my job (never to return) and became a full-time student at MVCC.

After I graduated from MVCC in 1970, with an AAS degree in retail business management, I enrolled at Utica College of Syracuse University and graduated cum laude in 1972 with a BS degree in economics.

Today, I am director of employment and training for Oneida County, and I have been a trustee of Mohawk Valley Community College since 1977. In the 19 years since leaving MVCC, I have served my community as a member of over 25 civic and community boards of directors.

All that I have become and will ever be, I owe to the education I received at Mohawk Valley Community College. MVCC showed me that a community college could be for everyone, including an average Black high school student.

David Mathis graduated from Mohawk Valley Community College in 1970 with a degree in retail business management and now lives in Utica, NY. He is director of employment and training for Oneida County and has served as chairman of the MVCC Board of Trustees and president of the Utica School Board simultaneously. He has also been chair of the Minority Issues Committee of the New York State School Board Association and executive committee member of the Associations of Boards of Trustees of SUNY Community Colleges.

Carmen Miranda-Jones
Salem Community College, NJ

I can recall, as far back as when I was a young student in Puerto Rico, always enjoying school more than anything else I did. I especially remember my dreams of attending the University of Puerto Rico on a full scholarship awarded to me for my high scholastic achievements. Then, in 1962, my parents decided to move to the U.S. mainland. I didn't want to go. I felt that learning a new language while adjusting to a new culture would affect my future educational goals. But I had no choice in the matter; I had to go.

As I had predicted, I suffered a setback. I had some problems learning the language, but I persevered and was able to get along quite well after a year. As I was preparing to finish high school, I sought the advice of the high school counselor to begin my college application plans. However, instead of advice, I received a devastating blow.

The counselor advised me not to attend college. He thought I would probably not make it because my "accent was still too heavy." According to him, college was "a different ballgame," and I would find college professors to be not as tolerant as high school teachers. He instead advised me to enroll in business school and take clerical courses, which would guarantee me a "nice, steady job." Guess what? I took his advice!

Yet, as time went on, and I kept successfully completing the clerical courses—which, by the way, were all in English—I began to wonder about the counselor and his wise advice. In 1980 I stopped wondering, challenged myself, and enrolled as a business administration student at Salem Community College.

As a student, I did far better than I had been led to believe I would, all thanks to the SCC professors who I found were very tolerant, understanding, and empathetic. They took me in. They helped me explore and recognize abilities that I was not even aware I had. They made me realize that I was indeed "college material," regardless of my accent. In fact, they encouraged me to keep my accent as a distinction of my language and culture.

SCC provided the hope and the right outlet for self-expression and -actualization. Its professors and staff became my facilitators. They provided me with an educational vehicle of which I took full command of steering. That vehicle led me onto a fantastic road paved with learning experiences, from which I am yet to recover! I am presently pursuing a master's degree in student-personnel counseling services at Glassboro State College. And I know it will not end there!

Learning is truly lifelong, especially when there are people who care about you. Thank you all for caring. Thanks also to my parents for their wise and timely decision. And to that high school counselor, wherever you are, thank you, too! The road has been long, but all learning is worthwhile!

Carmen Miranda-Jones graduated from Salem Community College in 1983 with a degree in business administration and now lives in Vineland, NJ. She is now an assistant manager of recruitment for the United States Census and is involved on a voluntary basis with developing and coordinating a homeless prevention program.

Jimmy F. Moore
Washtenaw Community College, MI

After completing the requirements for graduation from high school in 1959, I began to seek employment because I thought I would not be able to obtain a college education due to my parents' financial status. Scholarships were few, very competitive, and distributed in a biased fashion in those years. I found employment in a low-paying, unfulfilling position, got married, and started a family.

In the winter of 1961, I was appointed a police patrol officer with the Ypsilanti Police Department. Up until my appointment, few minorities ever held such positions in the city for any length of time.

After a few years, I began to feel the need for additional education, but I feared being in an environment competing with younger individuals. This feeling was pushed aside, where it remained dormant for years.

Socially, I found myself with people who were enrolled in community colleges. After many conversations with them and the untiring support of my family, I took that fearful step and enrolled also.

In the fall of 1976, I enrolled in the criminal justice program at Washtenaw Community College. To my surprise, I was welcomed in the setting by both students and professors. The intellectual experience and the creative encouragement of the professors nurtured my ambition to reach upwards in my profession as well as my desire for more information.

I graduated with honors after two short years of night classes at Washtenaw Community College, and without hesitation I enrolled in a senior college. Soon thereafter, I received a bachelor of science degree in criminal justice with honors from Madonna College in Livonia, MI.

My experiences at Washtenaw Community College prepared me for future professional goals. I received several promotions and in the spring of 1978 I became the first African-American ever to be appointed to the position of chief of police for the city of Ypsilanti. I retired from that position in the winter of 1986 after 25 years of law enforcement service.

I am now the owner of a successful full-service private investigation agency in Ann Arbor, MI.

All of this is owed to the "College of Hope" for without it, where would I be?

Jimmy F. Moore graduated from Washtenaw Community College in 1977 with a degree in criminal justice and now lives in Ypsilanti, MI. Currently a private investigator, he is a former policeman and retired chief of police in Ypsilanti. He conceptualized a master plan for the state of Michigan resulting in the adoption of its School Liaison Program.

Mary Musgrave
McLennan Community College, TX

In the spring of 1977 I decided to go to college. For many people that wouldn't have been a momentous decision, but at age 25 and after having worked for the state of Texas for five years, it was a decision that changed everything about my life. I was living in Austin and could have attended the University of Texas, but it was a world unto itself, and I'd never even set foot on the campus, let alone thought of attending such a massive institution. Instead, I chose to move back to my hometown of Waco, TX, and attend McLennan Community College in hopes of attaining an associate's degree.

My mother still denies thinking that I'd lost my senses when she heard I was quitting a perfectly good job to go back to school. Neither of my parents completed high school, and my older brother attended college for only a few semesters. In looking back on the decision to better myself, the primary reason that I am where I am today is due to my decision to attend a community college. I'd have been lost at a larger school and probably would never have received all the personal encouragement that MCC provided me.

The night to register for my first semester coincided with an infrequent Texas ice storm, and thus my first experience with college was late registration. I knew no one who was attending MCC at that time, so the selection of a class, "American History 1865-Present," was the start of my climb up the ladder to success. The instructors didn't care that I was an "older than average" student (which really wasn't true when "old" evening students and "young" day students were averaged out). They encouraged me to achieve my goal of getting an associate degree and then going on for higher goals, eventually receiving a master of library and information science degree from UT-Austin, that massive institution that five years earlier I would have been scared to attend. Through MCC and the community college environment, I learned that almost anything is possible.

I will be forever thankful to the faculty and staff at McLennan Community College for giving me the chance to become a librarian. They helped me to see past the self-doubts and gave me an excellent foundation for my work in public libraries. Today, I often have the chance to return some of the knowledge they instilled in me as I assist library patrons who are beginning or resuming their climb up the educational ladder. Now, it is my turn to pass on the encouragement that I received, and I hope that somewhere down the road those people will do the same for others who follow on their own path to success.

> *Mary Musgrave graduated in 1980 from McLennan Community College with a degree in sociology and is now living in DeSoto, TX. She is a librarian at DeSoto Public Library and vice president/president-elect of the University of Texas at Austin School of Library and Information Science Alumni Association.*

Donn M. Peevy
Gainesville College, GA

In May 1990 the oldest of my children, Shannon, graduated from high school. She is a good student, polite, well behaved, courteous, concerned, and caring for the well-being of her friends and fellow man. Her friends not only find her attractive, but also honest and dependable.

Though Shannon is a good student, a scholar she is not. Though she is warm and personable to her close friends and family, an extrovert she is not. Though she is courteous and well behaved, helpful and dependable, a leader she is not. And though she had good grades in high school, college is not in her immediate future plans.

Even though today I hold a bachelor's of science degree and juris doctor degree, practice law, and serve in the Georgia State Senate, where I chair a standing senate judiciary committee, 23 years ago at age 17 my high school profile read no differently from that of my daughter's today.

There are "born" scholars and "born" leaders, or so I've been told, and there are those exceptional people deserving of academic or athletic scholarships, and there are also those born to families financially capable of providing the resources for higher education at almost any institution of higher learning. I have no qualms with any of those who are so lucky or exceptional.

But there are some of us who are on a different track. Some may call it a slow track, some may call us late bloomers, but we march to the beat of a different drummer. Though in many ways we are very ordinary, I like to think of us as special.

We start late, we struggle longer, we work harder, we know we're further behind. It may be for a hundred different reasons; being painfully shy like Shannon (as was I); or being reared in a home where academics were not encouraged; or perhaps some physical or economic drawback. Whatever the reason, community colleges are special places for ordinary people seeking something more in life.

Community colleges are special because they are just that, colleges for the community, for people who live in the community, work in the community, and who are and will contribute to the community. Gainesville College was such a special place for me more than 20 years ago. I'm sure, should she so choose, it will be so for Shannon and all those other special people who realize extraordinary achievements from ordinary beginnings in special community colleges.

Donn M. Peevy graduated from Gainesville College in 1971 with a degree in liberal arts and now lives in Lawrenceville, GA. A former police officer and assistant to Congressman Ed Jenkins in Washington, Peevy is now a Georgia state senator and member of the board of directors of the Gwinnett County Chamber of Commerce. He also assisted with the fundraising campaign for the Gainesville College Foundation.

Michael F. Propes
Chemeketa Community College, OR

As a high school student, I had my first contact with Chemeketa Community College through trade classes offered in high school. Upon my high school graduation in 1973, I enrolled at Oregon State University in construction engineering under a football scholarship program. Due to injuries and three serious surgeries, I was unable to continue my studies.

I worked with my father operating our tree farm and heavy construction business. We were Polk County Tree Farmers of the Year for 1974.

When the opportunity arose for me to enter the forestry program at Chemeketa in 1978, the practical experience I had gained on our tree farm and construction business complemented the excellent forestry program there. I served as the student representative on Chemeketa's Forest Products Advisory Committee. Upon graduation in 1980, as the Outstanding Forestry Student, I was appointed to the Forest Products Advisory Committee as an industry representative.

The continued support and encouragement of the Chemeketa staff who nominated me for an Outstanding Young Man of America Award in 1981 have been instrumental in my success.

After graduation from Chemeketa I worked—simultaneously—as tree farm manager, forestry consultant, and broker of forest land. I also developed the forest products curriculum at our local high school and taught there for two years. For now, serving on the Polk County Board of Commissioners is a full-time job in itself.

Chemeketa also continued to play an important role in my family's life. In 1985 my wife, Pam, enrolled in the criminal justice program at Chemeketa. Upon graduation in 1987 she immediately went to work for the Oregon State Department of Corrections. Our children, Jason and Amy, have been supportive of our education. They help on the family Christmas tree farm and are looking forward to becoming Chemeketa students. Chemeketa Community College is truly a family affair.

This year as a Polk County Commissioner, I had the honor of taking part in the dedication of The Academy Building in Dallas, OR. This is a joint venture between Chemeketa and Polk County governments and was started when I was a member of the board of directors for Chemeketa. As a board member, I strongly supported this action. This historic building was saved from demolition and renovated to house the Chemeketa Dallas Center and offices for Polk County government—an example of local governments working together for the betterment of the community.

Yes, Chemeketa certainly has made it possible for me to be successful and to serve the public in my positions as a volunteer and as an elected official.

Thank you, Chemeketa Community College!

Michael F. Propes graduated from Chemeketa Community College in 1980 with a degree in forestry and now lives in Sheridan, OR. He is a Polk County commissioner and an active community and public service leader on economic development projects. His knowledge and dedication to his communities and schools have been acknowledged by several special awards.

Duane Schielke
Niagara County Community College, NY

Well-intentioned high school "guidance" led to my pursuit of an engineering degree. Statistically, good performance in the math and science college preparatory track and on standardized tests enables admission into universities of choice and provides for likely completion. Being what they are, however, there is always the down side of averages. Disenchanted with the rigors of extremely challenging and seemingly abstract studies, I left to seek more satisfying pursuits.

I enrolled part-time at Niagara County Community College, seeking a more meaningful outcome. However, leaving full-time study at the prime age of 19-and-a-half during the peak of the military draft meant otherwise; namely, induction into the Army. Ironically, those same math and science aptitudes qualified me for training in the art of gunnery and a subsequent tour of duty in field artillery in South Vietnam. A lifetime of worldly experience was compressed into a 14-month stint.

Upon return with humanitarian interests in hand, I re-enrolled at NCCC part-time evenings, flipping burgers full-time to fund day-to-day requirements in the adjustments of returning to the world. I found a student body composed of both first-time students and those returning after being first engaged in raising families, unskilled employment, or, like myself, military service. This breadth of backgrounds provided an enriching environment, nurtured by a faculty who acknowledged the dual responsibilities of many evening students.

I chose a career in human services after sampling courses along the way to an associate degree. Bachelor's and master's degrees followed, again while I was employed. From an entry-level position at a nonprofit social service agency, I advanced to chief executive officer and then associate executive of its state association, one of the largest nonprofit organizations in the country. NCCC not only fostered the formulation of meaningful career goals, but also served as an essential element in the foundation required to achieve them.

My affiliation with NCCC was rekindled along the way as I was invited to participate as an adviser and then as an instructor in a new rehabilitation services curriculum. While I was preparing graduates for community service jobs working with persons with disabilities, NCCC was a pioneer in the movement of community colleges to orient their missions to parallel the needs of their immediate communities.

For me, it afforded the rare opportunity to give back some of the benefits I once received. After seven years of teaching and helping the program flourish, I sadly departed to relocate, but with the comfort in knowing affairs were in good hands.

Duane Schielke graduated from Niagara County Community College in 1974 with a degree in liberal arts and now lives in Garrison, NY. He is the associate executive director of the United Cerebral Palsy Association of New York State and a member of the National Rehabilitation Association, the American Association of Mental Retardation, the Association of Medical Rehabilitation Directors and Coordinators, and the New York State Association of Community Residence Administrators.

Marilyn Shuey
Windward Community College, HI

Ialways wanted to go to college. Instead, I went to work as a clerk-typist right out of high school and married the following year. I was following the life plan laid out for girls and boys in the small Missouri town where I grew up. Girls were expected to get married by 21. Boys should get the college education if a family could afford it, because they would be supporting their own families early-1950s style. Twenty years later, I was to get the chance to go to college.

By 1970, my husband and I were settled in Hawaii after years of moving from one military base to another. Our son was born and I was content to leave a boring secretarial job for motherhood. Often during those years, I had the nagging feeling of being side-tracked, out of the mainstream. I was feeling the same as a lot of women my age who never had the chance to try our wings and learn some independence (a stage in life I believe no one should miss).

Windward Community College opened its doors in 1972. Here was my opportunity. Scared I couldn't make it, I signed up for a literature class—starting with something I liked, at least. I remember how I agonized over writing the papers. But by the end of the class, I earned my first "A" and loved every minute of it. The second semester, I took another literature course and added history. I had no thought of earning a degree; I was just putting one foot in front of the other.

WCC classes were demanding and exciting. Students ranged in age from 18 to 60 and represented many ethnic groups, making it all the more stimulating. Even more interesting, the instructors taught liberal arts with an egalitarian approach to the material, not the male-centered view I grew up learning.

Those of us who started college late in life were called the DARs, damned average-raisers. We were so eager to learn, to discover. Open enrollment let us in the door to find out what we could do, when we might have failed entry exams. Our patient instructors had to take us back to square one now and again, but we completed all requirements.

I earned my associate's degree in 1976, a bachelor's degree in political science (Phi Beta Kappa) in 1977, and finished a master's degree in political science in 1985. Along the way, I worked with community organizations and women's groups on legislation to open up opportunities for girls and women and to change laws and practices that allowed for discrimination based on gender. Today I'm working with the Hawaii State Commission on the Status of Women as a research analyst/statistician. I don't feel side-tracked anymore!

I wouldn't trade those WCC years for anything. Thanks to some wonderful teachers and a terrific program, I was able to accomplish the one thing I never thought I could.

> *Marilyn Shuey graduated from Windward Community College in 1976 with a degree in liberal arts/political science. After serving as the first public affairs coordinator for Planned Parenthood in the state of Hawaii, she has worked as a lobbyist on behalf of women's rights and is now employed as a research statistician with the Hawaii Commission of the Status on Women. Her efforts to change laws and build awareness of issues that affect women have resulted in many positive changes in the state.*

Mark D. Siljander
Triton College, IL

At nine years old, I developed a great interest and love for politics. John F. Kennedy's charisma had a profound influence in setting my life's vision. At that time, I set an ambitious course of goals for elected office, beginning at the local level and ending at the United States Congress.

Even though I had set these lofty goals, the zeal of youth overcame me, and classroom performance and achievement tests were unimportant during my high school years.

However, the last few months of my senior year in high school hit hard. If I was to succeed in my political vision, a college education was a must. Few alternatives were left. Triton College was not accredited at the time, but I felt it was my best opportunity to prove myself.

Triton College allowed me to come into my own. Uninvolved in extracurricular activities in high school, I became engulfed in college activities at Triton. In fact, it was a great honor for me to serve as the student representative on the Triton accreditation committee.

The emotional and intellectual support given to me by the administration and faculty was professional, caring, and made all the difference for an academically insecure young man.

After graduating with an associate's degree, I attended a state university in Michigan, went on to receive bachelor's and master's degrees in political science, and worked on a doctorate in education—focusing on community college administration. I was so grateful to Triton College that if my vision for elected office was not realized, I wanted to give something back to the system that gave me my chance at success.

As destiny unfolded, my vision was attained. At 21 I served four years in local government; at 25 I was elected and served four years in the Michigan State Legislature; at 29 I was elected to the United States House of Representatives and served three terms; and finally I served as one of the U.S. ambassadors to the United Nations—even exceeding the goals I had set in my youth.

Looking back, Triton College unquestionably had a profound influence in allowing me to participate in every legislative level an American could possibly serve in. In fact, one of my proudest moments occurred when I was asked to come back and share my appreciation and encouragement to the graduates as a Triton commencement speaker.

An old proverb states, "People without vision perish." I am very thankful Triton College was the catalyst that propelled me forth to achieve my life's vision.

Mark D. Siljander graduated from Triton College in 1971 with a degree in political science and now lives in Reston, VA. He is the president of a public affairs and import/export company. He has served in elected offices for 14 years, including the Michigan State Legislature and the U.S. Congress. He has also served as one of the appointed ambassadors to the United Nations.

Samuel F. Smith
Gainesville College, GA

Gainesville College was established in part to provide an opportunity for an affordable higher education for first-generation college students in north Georgia. Conceived upon the notion that only several hundred students would avail themselves of this opportunity, the college has continued that basic purpose, but for more than 2,300 students enrolled there today. I am pleased to be categorized as an alumnus based upon the academic and extracurricular activities that were available to me, for not only am I a first-generation college student, but I am also the first of 17 grandchildren to have attended college.

It has been 17 years since I was a student at Gainesville College. Since that time, I have actively contributed as a member of the board of trustees as well as the alumni association board of directors. During those years, I have come to appreciate the college's dedication to fellow north Georgians. Unlike schools that have evolved into major research institutions or those that have become forums for athletic events, Gainesville College has stayed the course—growing with the emergence of our geographical area and with the needs of baby boomers such as myself.

I have continued to appreciate the notion that first-generation college-bound students in the southern Appalachians could excel given an affordable opportunity. Even today, Gainesville College reaches beyond all expectations in providing financial assistance to virtually all students who demonstrate the need.

Perhaps it has been through my work as the chief of staff for a U.S. Congressman that my appreciation for Gainesville College has flourished. One-third of our staff resources is dedicated to helping individuals on a personal needs basis. Those, by and large, are the parents and grandparents of my college classmates who had no opportunity for college. They are individuals reaching retirement age and beyond who seek guidance in the myriad programs which might be available to them. They are citizens bettering themselves as we learned to do. My association with classmates from similar backgrounds at a time in our nation's history when much significance was thrust upon the South stays with me daily.

Could my parents have afforded other schools for me? Perhaps so, but they were pleased with my selection and the results thereof. And, as they had no educational opportunities whatsoever, they are proud of me today. Instilled in my generation is the attitude that we must provide for the next. Although it may seem trite (and Lincolnesque notwithstanding), Gainesville College remains of, by, and for the people.

> Samuel F. Smith graduated from Gainesville College in 1972 with a degree in political science and now lives in Gainesville, GA. He is the chief of staff for Congressman Ed Jenkins. Smith served on the Gainesville College Foundation Board of Trustees, Gainesville College Alumni Council, Gainesville Jaycees, Gainesville Vocational Rehabilitation Center, the Georgia Democratic Executive Committee, and such civic organizations as the United Way, Gainesville Booster Club, and the County Chamber of Commerce.

Mark Wellman
West Valley College, CA

I had injured my spinal cord in a mountaineering accident that left me minimal use of my legs. Being a very active person, I was concerned about what was going to happen to me and whether I had any chance to have some kind of career that would let me live in the mountains. I had heard of the park management program at West Valley College, which trained people for careers as park rangers and other jobs that involved the outdoors. I went to the office of the coordinator of the program, Tom Smith, and told him about my desire to have some kind of career that involved the outdoors. He told me that "there were lots of things that one can do confined to a wheelchair."

I entered the program in the spring of 1986. That spring I had the opportunity to even go on a High Sierra cross-country ski trip with the mountaineering class. I used a "ski" that was designed for me by a friend. It was the first extended period of time that I had been out of my wheelchair for three years! The summer of 1986, Smitty (Smith) talked to Bob Johnson, the district ranger for the Mather District, Yosemite National Park, about the possibility of employing me as a fee collector. That summer I was able to collect fees at the entrance station and the campgrounds in Yosemite.

I returned to the college program that fall and took classes in maintenance skills (including carpentry, plumbing, and trail work) and interpretation. I took a ranger skills class, learned how to ride a horse and pack a mule, and as a class project actually helped to design a whole access trail for the handicapped at Sanborn County Park. Sanborn Park is used by the college through a special cooperative agreement with Santa Clara County and is a 3,000-acre laboratory. The program at West Valley is unique and has a very well-rounded curriculum.

The next summer I was again hired by Yosemite to give guided walks and to operate the Happy Isles Interpretive Center. Today, I am a full-time ranger in the National Park Service in Yosemite. My duties include heading the handicapped access program, giving talks to visitors, meeting the public at the visitor center, and speaking to school groups.

I would like to be able to someday help design parks so that handicapped access is planned from the beginning.

I really think that when you go through two years with the same group of people, as I did at West Valley, you develop a closeness. The faculty and my peers were very supportive. I had a dream, and it has come true. I have always had the philosophy that it won't happen unless you go out and do it!

Mark Wellman graduated from West Valley College with a degree in park management and now lives in Yosemite, CA. He is a park ranger in charge of handicapped services for Yosemite National Park. Last July, after climbing El Capitan in Yosemite, he had a special audience with President Bush in the Oval Office. Wellman, a paraplegic, is an inspiration to all handicapped people in this country.

CHAPTER TEN

Education

David Anderson
Indian River Community College, FL

If someone had said to me back in 1959, "David, someday you will be a college administrator," I would have thought they were joking. After graduating from high school in 1959, I mowed lawns for a living and gave no thought whatsoever to attending college. Growing up the seventh of 12 children in a large, poor migrant family, I never considered college an option. Moreover, successful, fulfilling careers seemed as unattainable to me as touching a star. In life, it seldom happens that far-off stars drop into our laps; an opportunity, however, can make our dreams feasible. This opportunity came to me in 1961, when Leroy Floyd, president of Lincoln Junior College, recruited me.

After considerable deliberation with myself and my family about enrolling in college, I convinced myself to "go for it," and entered Lincoln Junior College when it opened its doors for the first time in 1961. A Black college, Lincoln later merged with Indian River Community College. I clearly remember that first year—all the reservations, doubts, and anticipations. This time of setting goals, facing new challenges, experiencing successes, and growing in quantum leaps took some adjustment. My sense of purpose grew stronger in the caring atmosphere created by the administration and faculty. Through their concern for my academic and social development, I truly felt I belonged.

My second year was a turning point in my life. Popular with both students and faculty, I was elected president of the student government association, thus beginning my political career. I completed my associate degree in two years, set my goals, and carefully mapped out my future. After flourishing in the community college environment, I set off with little trepidation for the university to earn a degree in political science. In 1973, I was awarded my doctorate in education by Florida Atlantic University.

Now, 28 years after entering Lincoln Junior College, I am vice president of student affairs at Indian River Community College and an elected member of a local public school board. Looking back with some amazement at my rewarding career as an educator, scholar, assistant principal, guidance counselor, associate dean of instruction, dean of students, and elected public official, I have only praise for the opportunities that abound in the community college system.

David Anderson graduated from Indian River Community College in 1963 with a degree in political science. He now lives in Stuart, FL, and is vice president of student affairs at Indian River Community College. In addition to his work with students and promoting minority education, he is active in the Martin County Drug Abuse Committee, the Council on Aging Board of Directors, the YMCA Board of Directors, and the Public Schools Advisory Committee. His contributions to the community and state were recognized when Phi Beta Sigma named him "Florida Man of the Year."

Robert E. Bahruth
Middlesex County College, NJ

In the spring of my senior year of high school I had still not applied to college and had no strong desire to do so. I was working in my father's business and earning a healthy salary. I didn't particularly love the hard physical labor (I was learning the trade of a mechanic, drove tow trucks, and changed and repaired thousands of flat tires), but I really hadn't been encouraged to go to college by my high school teachers or guidance counselors. Looking back now, I wonder if that was because my father was "only a mechanic."

Despite all this, at the urging of my mother, I applied to Middlesex County College just to give it a try. In August of 1968, I arrived on the MCC campus for freshman orientation. Since the college was in my hometown, I was surprised by numerous familiar faces. Many of my high school chums had decided to check out college as well. We jokingly referred to it as "13th grade."

I have to admit that I had many doubts about college, myself, and life in general. What I discovered at MCC was a pleasant surprise. While attending school those two years, I was stimulated intellectually, encouraged by some of my professors to be creative (many more than ever encouraged me in high school), and my ambitions in life were nurtured into specific goals. One professor who I will always remember and be grateful to is James Gallagher, the man who introduced me to linguistics and sparked my love of literature.

Looking back, I am now fully aware of the dramatic impact MCC had on my life. From the very first semester, I was hooked on college—an addiction that has lasted. Upon graduation in 1970, with an associate degree in liberal arts in hand, I was accepted to a small teachers' college in the foothills of the Smoky Mountains in Tennessee. MCC had inspired me to continue my education, but even more, I felt prepared to succeed. Prepared to make the big move away from my hometown, my family, and friends. Prepared to pursue greater academic goals as well. Two years later, I graduated with a baccalaureate in English literature and secondary education.

After 10 years of teaching, I returned to school, hungry for more. I got a master's in 1984 from the University of Texas at San Antonio where I was adopted by my mentor, Curt Hayes. I am now finishing my doctorate in applied linguistics and bilingual education at the University of Texas at Austin. I still like to work on my car now and then, changing the oil or tuning it up, but I realize how grateful I am that I don't do this for a living. Helping others to become successful educators is a much more rewarding endeavor for me. Thank you all at MCC. You provided the bridge I needed to find my calling in life. I know now that it was much more than thirteenth grade. I will always be thankful.

Robert E. Bahruth graduated from Middlesex County College, NJ, in 1970. He currently is an assistant professor of bilingual education and teacher education at Boise State University, ID. He has attended and taught in community colleges and considers both experiences invaluable. The author of several articles and chapters of books on applied linguistics, he is awaiting publication of his first book on bilingual education, Literacy Con Cariño: A Story of Migrant Children's Success, *which chronicles his years as a bilingual fifth-grade teacher in Texas.*

Dallas E. Barnes
Columbia Basin College, WA

It was shortly before high school graduation that I made the decision to attend Columbia Basin College. Although other college opportunities were available due to my athletic abilities, I felt a commitment to home. Limited finances and my own insecurities were important factors in my selection of CBC.

Looking back, without question, my experiences at the community college provided me with the confidence, support, and recognition necessary for me to aspire to the next level of achievement. In those two years, I competed on championship basketball and track teams, was honored as a campus celebrity, was recognized as an honor student, and graduated from college with an associate's degree.

The richest part of my community college experience, as it applies to my achievements in life and in my career, is not what it allowed me to do as a professional but in whom it has allowed me to be. Looking back, the community college experience gave me insight and an appreciation for the hopes, fears, and aspirations of individuals of different academic abilities, socioeconomic backgrounds, and religious preferences. Given these differences, as school-mates we shared in common the desire for self-improvement. This commonality allowed for cohesion and compassion to exist among us that resulted in friendships which, for me, have lasted to this day.

Many teachers contributed significantly to my education, particularly, Lenor Van Sant, zoologist, and James Beecroft, mathematician. Both took the initiative and made the extra effort to extend to me the guidance and education I needed, as opposed to just the subject matter they taught so well in their classrooms. From them, I learned about human potential and how it could be unlocked by educators who are willing to take the initiative and make the extra effort. For me, these teachers' initiatives and extra efforts made a difference. Today as a university official, I consider it not to be my professional prerogative to make the extra efforts for others but, having been a recipient, my duty.

Dallas E. Barnes graduated from Columbia Basin College in 1961 with a degree in social sciences. He now lives in Pullman, WA, and is director of academic development programs, Student Advising and Learning Center at Washington State University. He has organized and established an academic retention program for athletes on scholarship, developed and instituted an educational opportunities program for single teenaged parents at the university level, and established a graduate students' internship program for training learning center specialists.

The community college experience has helped me to learn that there is little difference in the things to which people aspire, when all else is equal. The difference in their attainment rests solely on the opportunities they have to achieve their aspirations. My current position at Washington State University permits me to be a primary advocate for those students whose life chances made it difficult for them to have the opportunity for a higher education. Many have returned to express gratitude for the extra efforts, efforts that are products of the values firmly rooted in my experiences at Columbia Basin College.

Phillip O. Barry
Moraine Valley Community College, IL

In 1972, I had recently completed a two-year AMA-approved program in radiologic technology and had hopes of entering Chicago Medical School to earn a bachelor's degree in radiological sciences. The prerequisite was at least 32 hours of general college studies. I needed a college that was progressive enough to realize that my training and accomplishments as a "nontraditional" student were significant. Multiple colleges would not accept my training as collegiate-level and would not recognize that I was nationally registered. These colleges continued to emphasize my high school grades. Finally, someone suggested a community college. After some research I learned that Moraine Valley Community College was one of the newest (five years old) and most progressive in the area.

I drove to the college; it happened to be a registration period. I walked up to someone from student services and informed him I was interested in enrolling. I'll never forget the next statement. I told him, "My high school grades were not the greatest, but I've just graduated from a two-year radiologic technology program." The staff member's response was, "Don't worry about the past; we want to help you with your future, but we will evaluate your recent training." That was the first time someone in education showed a genuine interest in my future. I enrolled that day and started the long path to lifelong learning.

I remember the scheduling of classes and weekend human potential classes that seemed to be designed for students like me, working full-time and raising a family. Not only did the scheduling fit my needs, but the faculty were unique—unique in the fact that they were caring, flexible, and excited about what they were doing.

I knew then what I was experiencing was different from the educational institutions I had known or heard of before.

Moraine Valley Community College, its staff, and faculty practiced the philosophy of a true community college and allowed me to develop my abilities and succeed in a college environment that enabled me to see the advantages of a college education. This experience allowed me to go on to obtain the associate's, bachelor's, master's, and doctorate degrees.

My positive experiences at Moraine Valley Community College have taken me from student to faculty member, program director, associate dean, and dean. Today, as a community college president, I can say "I was born and raised in the community college."

Phillip O. Barry graduated from Moraine Valley Community College in 1973 with an associate's degree in radiologic technology. He now lives in Woodstown, NJ, and is president of Salem Community College. He went on to obtain his bachelor's, master's, and doctorate degrees, which took him from student to faculty member, program director, associate dean, dean, and president. He has contributed a great deal to the advancement of education through his professional activities.

Yvonne P. Bergland
San Diego Mesa College, CA

When I moved to the United States from Canada I knew that it was supposed to be the land of opportunity, but to my surprise Mesa College opened more doors and gave me more choices in life than I had ever imagined!

My career path was somewhat atypical. After I graduated from the Edmonton General Hospital School for Medical Record Librarians in 1965, I came to the U.S. and discovered I needed a bachelor's degree in order to practice my profession here. In the meantime, as a part-time job, I taught medical records courses in the evening. This preview was all I needed to realize that I wanted to be a full-time teacher. Since this required further education, I decided to begin at Mesa College—where else! I was actually more nervous about becoming a student than I was when I became a teacher at Mesa College. It had been 12 years since my formal education days. Fortunately, a counselor who worked with reentry women encouraged and guided me through the requirements when I transferred to a four-year institution.

During my Mesa years, my instructors were constantly challenging me to reach beyond what I thought were my limits. When I graduated with honors in 1978, I was eager and ready to continue my education at San Diego State University. Motivated by a love of learning that was inspired by people at Mesa College, I decided to finish my master's degree as well. By the time I earned my master's, I was a full-time contract instructor at Mesa College and director of the medical record technology program there.

Still filled with a desire for learning, I went on to earn a doctorate at Ohio State with an emphasis in adult and higher education administration. Shortly after I returned to my position at Mesa College, I was offered the position of dean of the School of Health and Natural Sciences with responsibility for four departments and 14 disciplines.

Going from student to dean in 12 years is a quantum leap I never would have made without the initial boost and continual support I received from my friends and colleagues at Mesa College. It's truly the college of opportunity!

Yvonne P. Bergland graduated from San Diego Mesa College in 1978 with an associate in science degree. She now lives in Santee, CA, and is the dean of the School of Health and Natural Sciences at Mesa College. She devotes much of her leisure time to helping her profession—medical record technology. Her colleagues have recognized her excellence and abilities by electing her to key posts locally and nationally. Last year she was named the outstanding vocational instructor at Mesa College. This year she received the Professional Achievement Award from the California Medical Record Association.

Lynn C. Black
Central Community College, NE

When I entered college in the 1960s my career goals were somewhat in limbo. As a result of working in a plastics injection molding company, I was driving a new vehicle and had money in the bank. Attending college meant I would have to give up the job and the car, but I decided to make the sacrifices because of the encouragement of my parents and friends.

My first year at a four-year college was very discouraging, and I decided not to return. I was, however, still interested in pursuing some form of higher education and I was fortunate to discover Central Community College. Through the dedicated efforts of their faculty and staff I was able to determine a career goal in the field of education, ultimately hoping to return to a community college environment.

After completing an associate degree in business administration, I returned to a four-year college to obtain a baccalaureate in business education. I taught in the public schools for nine years, including evening classes for the community college, while continuing graduate work. I was able to earn a master's degree in vocational education, a specialist degree in educational administration, and a doctorate in adult and continuing education.

I was able to return, not only to the community college system but to Central Community College, to give back to other students what had been given to me—an inspiration to learn and succeed. My first position was as chair of the Business Department at the Hastings Campus, then associate dean of students on the Grand Island Campus. Currently I am dean of student services.

Through my experiences in the field of education, I have become involved in a variety of national, state, and local organizations as well as religious, charitable, and community groups. Some of these organizations include: Phi Delta Kappa, Nebraska Community College Student Personnel Administrators, Chamber of Commerce committees, Rotary, Elks, and the Knights of Columbus.

Education at the community college level gave me the skills and knowledge to pursue a goal—a goal I would not have reached without that beginning. I cannot think of a more rewarding educational career than working with a dedicated staff whose priority is to provide the education necessary to assist adults in their educational goals, from early entry high school students to those well into their retirement years. The educational offerings on and off campus, service to business and industry, and the dedication of the community college employee in assisting the total community is quite overwhelming. Without a doubt, community colleges, and especially Central Community College, are the best kept secrets in the total realm of higher education.

> *Lynn C. Black graduated from Central Community College in 1969 with an associate's degree in business administration. He now lives in Grand Island, NE, and is the dean of student services at CCC. Through his experiences in the field of education, he has become involved in a variety of national, state, and local organizations, including Phi Delta Kappa, Nebraska Community College Student Personnel Administrators, the Chamber of Commerce, and the Rotary Club.*

Robert L. Brown
Rend Lake College, IL

I guess I was actually inspired by my second grade teacher to become a teacher. She was my idol; she cared; and she provided a positive motivation for me to do my best.

In my family, my grandmother and my aunt were my lifeline to the future. My grandmother always said for me to learn to get along with everybody and to look for the "good" in others. One of her main thoughts was to "make something out of yourself."

My goal of wanting to be a teacher and return to Mt. Vernon to teach was ever present. Due to our financial situation, Mt. Vernon Community College (Rend Lake College) was my only consideration. The tuition was $10 per semester. Looking back, we had to make some sacrifices to get it.

Once I entered college, Betty Ann Ward provided me with guidance. She took me under her wing and made me feel like I counted. Other faculty members provided the one-on-one type relationships that helped me succeed. If I started to have problems in English composition, the teacher, Todd Oliver, would take me aside and offer suggestions for improvement. The caring, dedicated teachers helped to make a difference for me.

The students were friendly. However, I was always taught that to make a friend you had to be a friend. I made it a point to speak to other students and they spoke back. I was involved in many activities like the Student Education Association with Imogene Book. The administrator, Howard Rawlinson, had the door open to students to make us feel that we counted.

My years at the college made a difference in my life. My confidence in my academic abilities improved; my self-esteem improved greatly; and my continued willingness to make something out of myself was fed with a commitment to be a successful teacher.

After completing my associate degree at Rend Lake College, I graduated from Southern Illinois University in 1968 with a bachelor's degree in social studies, and in 1973 with a master's degree in guidance and educational psychology. I am currently a guidance counselor at the Mt. Vernon Township High School in Mt. Vernon, IL.

Since I am in the public eye as a school counselor, I have been involved in many church, school, and community related activities. I really like my job and enjoy the people I work with at school. I owe it to my family and to my teachers from the elementary level through the junior college level who provided the "I know you can do it" attitude.

Robert L. Brown graduated from Rend Lake College in 1966 with an AA degree. He now lives in Mt. Vernon, IL, and is counselor at Mt. Vernon Township High School. He is a member of the Jefferson County Credit Union Board, the NAACP, the Mt. Vernon Education Association, the Illinois Education Association, and the Illinois Association for Counseling and Development. He is a former board member of the Lutheran Children and Family Services and former co-chairman of the House-to-House Campaign for the American Cancer Society.

James E. Bruno
Mt. San Antonio College, CA

Without a doubt, the community college experience was the most important educational and social institution in my academic development. I was blessed with magnificent parents who also happened to be poor. I attended Mt. San Antonio Community College with my twin sister. At Mt. SAC we both received a first-rate education and upon graduation successfully transferred to a four year institution. My mother, upon attaining her high school equivalency, also attended and graduated from Mt. SAC. The community colleges in general, and Mt. SAC in particular, have a very special place in our family.

I am now a full professor at UCLA—the very institution that I transferred to upon my graduation from Mt. SAC. As a teacher I see countless thousands of undergraduate students being educated in large impersonal lecture halls. Most have teachers who are more interested and bonded to their research than to teaching. Many of these students also have poor academic preparation, and I feel that most would have been much better served by attending a community college before transferring to a four-year school.

The important initial bonding of a student to educational values is significantly enriched when the student feels that he or she is the centerpiece of the instructional program. The community colleges have always made the student the centerpiece of their instructional programs. At least for me, I always felt this special relationship and greatly benefited as a result.

I distinctly remember my classes at Mt. SAC. The small class sizes, the quality teachers and, of course, the incredibly low cost. I have many fond memories of this important phase of my development. Outside of my classes at Mt. SAC, there were many special interest groups and athletic programs. I am still active in many of these areas. In fact, my future occupational interest was due, in large part, to my participation in one of the Mt. SAC special interest groups having to do with the teaching profession.

Finally, as the decline in the quality of American elementary and secondary education continues unabated and the cost for attending four-year college institutions escalates, the future role of the community college will be significantly enhanced. The community college will be called upon to deliver on the promise of equal educational opportunity for the vast majority of citizens. It will also become the largest provider of educational services in the basic skill areas of mathematics, language, and reading. These responsibilities will be in addition to its traditional role of vocational and pre-four-year college preparation.

The community college was for me and will continue to be for students in the future, the best educational bargain in the world.

> *James E. Bruno graduated from Mt. San Antonio College in 1970 with a degree in liberal arts. He now lives in Los Angeles, CA, and is professor of education at UCLA.*

Charlotte Campbell
Gadsden State Community College, AL

Thank goodness for counselors like Deborah Beverly at Gadsden State Community College. Every time I see her or send a new student to her for college advisement, I thank her. I can truthfully say that if it had not been for her encouragement, I would not have enrolled in college. Sad to report, she is the only person in education who even made the suggestion that I attend college.

After graduating from high school, I worked as a secretary, a cosmetologist, and a sales representative. I earned a substantial income but was never totally satisfied with my employment. Twelve years after graduation I began college.

I must admit that when I first enrolled as a freshman I was scared to death. However, Gadsden State provided just the right challenge for me. The instructors were personable and eager to help. I could hardly believe it when I completed the associate in science degree.

Next came the distant call of the four-year university. I answered the call at Jacksonville State University in Jacksonville, AL, and earned a bachelor of science degree in early childhood education. While at JSU, I completed a master's degree in education.

After teaching four years in an elementary school, I was promoted to a supervisory role in the Central Office. Of course, this called for—you guessed it—more education. In 1989 I completed an education specialist degree in administration. I am currently coordinator of special projects in the Gadsden City School System in Gadsden, AL.

In addition to my role in the Gadsden City School System, I have taught as an instructor of early childhood education for Jacksonville State University in Jacksonville, AL.

My involvement in professional organizations and in the community is very important to me. I am a member of many organizations including Delta Kappa Gamma, Phi Delta Kappa, and the Association for Supervision and Curriculum. I am also a member of the Chamber of Commerce and a volunteer for United Way.

Recently, someone asked if I planned to continue my education. At this point in my career I have to really stop and think about that question. Who knows? Once these community colleges get you hooked on education, I think the priority is probably education for a lifetime.

Charlotte Campbell graduated from Gadsden State Community College in 1979 with a degree in general studies. She now lives in Gadsden, AL, and is coordinator of special programs for the Gadsden City Schools. After accidentally discovering her learning style in an audio-tutorial job, she went on to copyright two systems geared to auditory learners. Despite her hectic work schedule, she is an active member of the Gadsden/Etowah Chamber of Commerce and the United Way. She chairs several committees and chaired the county-wide summer festival "Riverfest '90."

Kathy Carlton
J. Sargeant Reynolds Community College, VA

In 1964 I graduated from Richmond public schools and entered Longwood College, the goal for which I had prepared. But, having no career expectations, I floundered. Dropping out after a year, I completed a secretarial course at a Richmond business school and entered the work force. During the next 15 years my secretarial positions were varied but unchallenging.

Significant events included the accidental death of my father, marriage and divorce, and the birth of my precious daughter, Misti. An important month was July in 1982 when I decided to try and free my life of addiction to alcohol that had plagued my immediate family and my marriage. A frightening step in my recovery from the ravages of alcohol was finding enough self-esteem to enroll in one class at J. Sargeant Reynolds Community College. Though concentration was difficult and frustrating, the class brought me my first "A" in years.

Then I became an assistant bookkeeper, running computerized payrolls and being introduced to simple accounting responsibilities.

In 1987 I re-entered JSRCC in the accounting curriculum while working full time. Gaining confidence from another "A," I went to a counselor, changing my major to business administration, a transfer curriculum.

While taking two classes in the next semester, I was confronted by a twist of fate that altered my life dramatically. Eleven-year-old Misti was riding with a friend's mother when their car was struck by a freight train. While Misti lay in a coma, doctors told me if she lived, she would never be the same person she had been.

As Misti recovered and began functioning normally again, I returned to work. Months later my employer closed his firm, and I had the opportunity to attend JSRCC full time on scholarships for one year, spending my savings for living expenses. Eventually my degree work was completed while I was again working full time.

Everyone at the college was extremely supportive and encouraged me to set higher goals for my education. It was a surprise and honor to be selected a Virginia Scholar - one of only five community college students in the state. A national accounting society presented me with a scholarship from among 2,200 applications nationwide. The University of Richmond Business School offered me funding to complete my tuition package for 1989-90, and I am thrilled to be a junior accounting major at that respected institution.

Thanks to JSRCC, I now have enough self respect to believe anything is possible, if we are willing to work for our dreams.

> *Kathy Carlton graduated from J. Sargeant Reynolds Community College in 1989 with a degree in business administration. She now lives in Mechanicsville, VA, and is attending the University of Richmond Business School full-time as an accounting major. She has overcome drug and alcohol addiction and her daughter's near-fatal accident and gone on to excel in school. In 1989 she was selected a Virginia Scholar— one of only five community college students in the state named to the list of honorees.*

Mary Ann Carnegie
Washtenaw Community College, MI

When I finally got serious about going to college, I was already earning the kind of salary most new college grads hope for. I was also a full-time wife and mother. Still, I couldn't assuage the guilt I felt living in a city where two universities and two colleges were only minutes from my home. So, prompted by guilt and a deep-seated love of learning, I enrolled at Washtenaw Community College.

My husband drove me to the college and practically held my hand through the registration process. Soon, however, I was rushing home from work to play Mom before going off to class.

I loved it! It never occurred to me that what I was doing was hard. It was actually relaxing to sit in class and listen to an instructor or interact with other students. The classes all seemed interesting to me, especially the ones that were the most challenging.

I realize now that I looked upon the instructors with something close to reverence. Many of them secretly became my role models. I admired their self-confidence and what seemed like limitless knowledge. I was convinced that a college education could do the same for me.

It has! With my associate degree completed, I immediately transferred to a nearby university where I completed a baccalaureate degree.

Now, I have returned to Washtenaw Community College where I am employed as coordinator of community and alumni relations. Promoting the college's mission first as a student and now as an employee enables me to represent the college in a way that is genuine and truly fun.

From time to time, students approach me in a way that reminds me of the shy student I once was. I know then that I have a responsibility as a role model and I take that responsibility seriously. It's wonderful to give back to an institution that gave so much to me.

Mary Ann Carnegie graduated from Washtenaw Community College in 1980 with a degree in general studies. She now lives in Ann Arbor, MI, and is the coordinator of community and alumni relations at Washtenaw Community College. A caseworker for the State of Michigan Department of Social Services, she previously worked as a counselor for a private rehabilitation company teaching individuals how to find employment.

Gregory A. Carrington
LaGuardia Community College, NY

When I went to LaGuardia, I had a "street attitude." LaGuardia was a life-saver for me. It taught me to get myself organized. It gave me the opportunity to interact with many different types of people. I had many good professors who gave me a lot of personal attention and helped me to clarify my values. If I hadn't had the LaGuardia experience, I'd probably be on a street corner or in jail today.

Instead, today I am married with two wonderful sons. I continued my education and am now in a master's program in special education. I have a marvelous career. My main focus has been to serve the youth of my community. I am a counselor/ teacher in a special New York City school program. After hours, I work as a volunteer for the after-school center and provide tutoring and counseling to the students. I also am a scoutmaster for the Boy Scouts, coach two Little League teams, and serve as a youth coordinator for a local poverty program where I help coordinate fund raisers, community-based referral services, and drug prevention programs.

I have a full, rich life. I know what my priorities and purposes are. I attribute this to LaGuardia which is a place that was my stepping stone. It gave me direction, a sense of organization, and a get-involved attitude.

Recently, at the college's graduation ceremony, I received the Alumni Association's first award recognizing a "graduate who has successfully achieved his or her career goals and exemplifies the spirit of LaGuardia through demonstrated commitment to his or her career, community or school." I was deeply honored, especially because I feel I owe it all to LaGuardia.

Gregory A. Carrington graduated from LaGuardia Community College in 1976 with a degree in human services. He now lives in Queens, NY, and is counselor/ teacher in a special New York City school program. Despite a chronic kidney disease that requires him to have regular dialysis treatments, he is actively involved with the youth of his community. LaGuardia Community College honored him as a "graduate who exemplifies the spirit of LaGuardia through demonstrated commitment to his career, community, and school."

Winslow Lee Carter

Chowan College, NC

Attending college had been a dream as early as I could remember. No one in my family had finished high school and there was little commitment to higher education. College seemed like an impossible dream. Fortunately a teacher in my high school encouraged me to give college a try. He and the school counselor said they knew just the college for me, Chowan College.

In August of 1969 I left for Chowan, 130 miles away, the farthest I had ever been from home. It was a scary but exciting experience for me. The friendliness of the faculty and staff calmed my fears. I began attending classes with very little confidence in my academic ability. I gave it all I had. The first examination was in my religion class. I felt sure I had done poorly. When the professor returned the test, she said the grades ranged from 40 to 99. I swallowed a lump in my throat. I was absolutely stunned when my paper was handed to me and I had made the 99. What that teacher and that grade did for me was a turning point in my life. For the first time in my life I began to believe in myself.

Gradually I began to realize a potential within me that I had never seen before. For me college became a sheer joy. I loved every minute of it. I was amazed at myself. I was in two plays, Spanish club, men's council and made the dean's list. My mind was stretched as I grasped new truth and the whole world became an experience from which to live and learn. Chowan became the springboard to what has been, to date, an exciting and rewarding life.

Chowan gave me excellent preparation as I transferred to a senior institution and received my bachelor's and master's degrees. It opened the door to another dream I had always had when I heard a young lady speak about living in another culture. That dream came to fruition when I graduated with my bachelor's and went to teach school for two years in Kenya, Africa.

The world Chowan opened up to me helped me realize just what a mission and opportunity education can be. What better way could I repay the debt I owe than to give my life to the vocation of education, helping others catch the spirit of learning and finding truth.

Today as director of admissions at Chowan College it is a great opportunity to see what I experienced being repeated over and over again. It is wonderful to have a part and feel that in a small way I am paying a little on the debt of what Chowan did for me.

Winslow Lee Carter graduated from Chowan College in 1971 with a degree in liberal arts. He now lives in Roanoke Rapids, NC, and is the director of admissions at Chowan College. After graduating with his BA, he went to teach school in Kenya, Africa, for two years. He is an active member of the First Baptist Church in Roanoke Rapids and a former member of Rotary International.

Carol Churchill
Southwestern Michigan College, MI

I sat beside a graceful blond, fully aware that she had probably paid more for her jeans than I had paid for my first car. I would not have been in that university classroom at all, a thirtysomething wife and mother, had it not been for my long ago experience at Southwestern Michigan College.

Seventeen years previously, I had earned an associate degree at SMC. During my two years there, I thrived in an atmosphere characterized by the college's motto, "Excellence With a Personal Touch." The brand-new community college was staffed by eager instructors who both nurtured and challenged us. Under their tutelage, I developed a repertoire of academic capabilities that enabled me to graduate summa cum laude.

After graduation, I married and began what was, in retrospect, an idyll: beautiful children and a bountiful life on a small farm. We didn't notice the fragility of our pleasant bubble until the agricultural economy actually burst. Prices plunged; interest rose. For the first time, I felt fear. I knew that I needed something apart from the farm, something that would remain steadfast in the whimsical winds of the weather or the political climate. Despite a brain that felt as dusty as my set of Shakespearean plays, I decided to return to college for a bachelor's degree.

Many times during the registration process, my shaky confidence wavered. My resources were very limited, but all doors seemed closed at the financial aid office until I mentioned that I was a somewhat belated transfer student from a community college. Within a week, I was the recipient of a two-year, full tuition scholarship to the university, based on my academic performance at SMC 17 years before!

With one barrier overcome, I began wondering if all my credits would transfer. My worries were unfounded; in fact, enough requirements had changed over the years that I would graduate with a double major. By the end of my first month, I felt as though I had never been away from the academic world. Two years flew by, and I graduated with a 4.0 GPA, success directly attributable to the academic foundation that had been laid so many years before at SMC. Filled with euphoria, I eagerly began my job search.

Ironically, my search led me home again; I am now a placement counselor at Southwestern Michigan College. For my children, the circle that led me here again testifies to the value of education. For others, the circle demonstrates the endless possibilities that can begin at the college in their own backyards. For myself, even though I am still working on my master's degree, I have been able to use my own experiences to help clients conquer their fears and begin anew in either college or the workplace. Once again, I am thriving, both personally and professionally, in the caring environment of a community college.

> *Carol Churchill graduated from Southwestern Michigan College in 1968 with an associate degree in English. She now lives in Marcellus, MI, and is a placement counselor at Southwestern Michigan Community College. In addition to working on her master's degree, she is very involved in the agriculture community as the wife of a farmer and in school activities as the mother of three.*

Duncan Coons
Northern Oklahoma College, OK

My first memories of Northern Oklahoma College were the football games that I attended with my father in the early sixties. The college had an enrollment of 500-600 students and was located in Tonkawa where it still serves north central Oklahoma. My parents' expectations were that I would continue education beyond high school, but Northern wasn't chosen until my senior year. Entering Northern, I met Edwin Vineyard, Northern's president. Our acquaintance would later represent much more than our first year student/administrator association.

After three semesters, and patience on the part of Northern's staff, I chose music education for a major. The influence of Bill Heilmann, fine arts chair, helped guide my future. My Native American status fit in well with the homogeneous population of 1,000 on campus. I enjoyed participation in many campus activities while serving as drum major for the Northern band. Though not earning my associate degree, I transferred hours and valuable experience to a state university, earning my bachelor of music education degree in 1971, following military service.

Not expecting to be associated with Northern after leaving, I was able to schedule their famous "Roustabouts" vocal organization into schools where I taught. The next association came when teaching in Northern Oklahoma during the late seventies. I again found myself on the Northern campus. Three of the four years I had taught in Northern Oklahoma I served as chairman of an honors band which the college sponsored. The renewal of friendships and assistance by the Northern staff was refreshing and quite welcome.

Moving to central Oklahoma allowed me to complete the principal and superintendent provisional certification without commuting. A superintendent position opened at Braman, OK, not far from my home town and 20 miles from Northern! I confided to President Vineyard that I wanted to earn the associate degree from Northern that I missed some 20 years before. The cherished degree was received at the conclusion of the 1988 spring semester. Would I be on campus again? I took a course at Northern to complete administrative certification requirements during the fall of 1988.

There was no reason to return to Northern. Or was there? The North Central Oklahoma Administrators' Association was formed in the fall of 1988 and I was elected president. Northern Oklahoma College, of course, volunteered to host meetings! I know that I will return often to visit the busy campus of Northern Oklahoma College.

Duncan Coons graduated from Northern Oklahoma College in 1988 with a degree in music. He now lives in Braman, OK, where he is the Superintendent of Schools. Mr. Coons did not complete all degree requirements before transferring in 1967. He returned after completing a graduate degree and made a special request because he wanted the degree from NOC. He has been a highly successful music director, winning honors for his bands.

Sharon J. Cotman
Thomas Nelson Community College, VA

In high school I was enrolled in typing and shorthand classes. Due to limited financial resources, I planned to be a secretary when I graduated and did not have any plans for college. During the summer prior to my senior year, I was selected to participate in a pilot secretarial science cooperative education program with Thomas Nelson Community College, NASA, and local high schools. It was an excellent opportunity to gain valuable work experience while going to school. Upon my high school graduation, I continued working at NASA in the secretarial field and attended college too.

During my freshman year, the instructors stimulated me; the combination of work and school was very challenging, yet rewarding. It was exciting to be able to transfer the knowledge and concepts presented at school into practical application at work. Through this experience, I gained insight into various professions and was motivated by a counselor to pursue a dual track degree to have the option of transferring to a four-year college.

TNCC was my second home; I liked it so much that I did not want to leave. After graduation, I worked full-time as a secretary but continued to attend night classes for the next year. I did eventually move on, though. In Fall 1977, I was accepted at a four-year state college, and after two years of night classes received a baccalaureate in business administration.

The combination of education and six years of government secretarial experience proved to be assets as I changed careers to become a business applications computer programmer. From 1979 to 1984, I progressed to programmer analyst and through the years trained many secretaries and clerks to use computer systems that I had been involved in developing. In Fall 1981, I began night school again and two years later received my MBA.

The following year, 1984, I was selected to be an assistant professor in the business division at TNCC. Now, I was on the opposite side of the desk and became a peer to the group of professors who taught and inspired me. After five years of teaching at TNCC, I am sure I made the right choice to attend there and to return as an assistant professor a decade later.

Without TNCC, I doubt that I would have pursued a college education. The foundation was laid at that college for my future potential in terms of career and educational endeavors. The most rewarding aspect of teaching at my alma mater, TNCC, is making an effective contribution by assisting and perhaps inspiring other students to become successful. I am thankful the community college was there when I needed it.

Sharon J. Cotman graduated from Thomas Nelson Community College in 1976 with an AAS degree in social science and is now living in Grafton, VA. She is an assistant professor of computer information systems at TNCC. Also a faculty advisor and church volunteer, she conducted a senior citizen computer literacy workshop in fall 1989 and helped to develop a microcomputer systems business certificate program now offered by Thomas Nelson.

Charles R. Crumly
Southwestern College, CA

It so often seems that momentous decisions are not perceived as such, while the seemingly unimportant decisions are those that change a person's life. So it was when I chose to attend Southwestern College. I had no particular plans, no special goals, only a few academic interests, and no money for tuition. My time in high school was spent with cross country and track and field. I was young for my age - the quintessential late bloomer - 6'3" and woefully thin. I was unconcerned with academics, and my grade point average showed a lack of self-discipline and maturity. My only assets were curiosity and open mindedness. The fact is, with my record, I probably would not have been admitted to any decent four-year college.

My classwork at Southwestern was mediocre. I did well in the classes I liked (my most memorable class was Invertebrate Zoology). My star did not suddenly burn brightly after arriving at Southwestern. However, with hard work and a higher level of competition, my running was improving. For me, this was more important.

Aside from sports, my academic interests remained unfocused. I had thought I might be a physical education major - teach biology and coach the cross-country and track teams. My first theory of P.E. class was a deadly bore. Then I thought I might go into anthropology. My imaginings centered on important excavations, uncovering Man's history, pith helmets, and intrigue. Next I thought that pottery was for me; I would be throwing pots, working with clay, creating form and beauty from nothingness. However, through all these side trips, I kept returning to biology and to running. My experiences at Southwestern were a part of this exploration, but certainly not all of it. Eventually, I settled on biology and zoology. I also ran in the New York City Marathon, finishing quite well in just over three hours.

My memories of Southwestern are not vivid, and my time spent there did not especially prepare me for my eventual goals. Nevertheless, Southwestern and my athletic experience there gave me the confidence to try. Though my memories are fading, the time spent was valuable, for it prepared me in many intangible ways for an unanticipated future. I suspect that my experience is not unusual. Many of us stumble, as much by accident as planning, into life. Southwestern gave me the chance to stumble. Thanks.

Charles R. Crumly graduated from Southwestern College in 1971 with a degree in biology and is now living in San Diego, CA. He is a zoologist, scientist, and teacher at both Southwestern College and San Diego State University. He is also a research assistant for the Bureau of Land Management, an Audubon correspondent for the American Museum of Natural History, a visiting curatorial associate at Harvard University, and a consultant to the National Wildlife Federation.

Wayne G. Deahl

Eastern Wyoming College, WY

I didn't want to go to a community college, but since there was one in town and money was tight, that's where I ended up. And I'm glad it happened that way.

After graduating from high school, I was convinced that I had no desire to attend the "high school on the hill," since my grades and test scores indicated to me that I was meant for bigger and better things. But the reality was that my education career was not ready for anything else. If it had not been for understanding and caring instructors, for small class sizes, and for individualized attention, the first year of my college career would have been an entire waste.

As it was, EWC provided an outlet for my energy in many ways. Admittedly, I spent little of it in the classroom, but rather in theater, journalism, music, and the like. Thus, I was able to enjoy college in a way that a university would not have allowed me. My grades satisfied minimal requirements, and then only because instructors were willing to challenge me and work with my attitude toward education.

And what a joy it was to return to school some three years later and discover that I really enjoyed learning. Many of the same instructors were at EWC, and they had no stereotypes concerning my abilities. They were willing to allow me to grow, and I can only express extreme gratitude for their efforts. The quality of education and the quality of the experience at a community college cannot be duplicated elsewhere.

There is no departmentalization at EWC. The student is free to associate with the rest of the student body and not become involved with the cliques of various schools that one finds at the university level.

I must say thanks to those at EWC who made my learning experience there a vital one; one that set me on the path to career success. If I had been able to attend the university I had chosen, I'm sure my life would have been much less successful and enjoyable.

> *Wayne G. Deahl graduated from Eastern Wyoming College in 1974 with a degree in secondary education/theater and is now living in Torrington, WY. He is an instructor of English and theater at Eastern Wyoming College and was recently appointed by the governor of Wyoming to serve on the Agenda 2000 Advisory Committee for the University of Wyoming. He was also a charter member of an advisory board for the Laramie public schools and a board member of Big Brothers/Big Sisters of Southeast Wyoming.*

Linda Dickman
Burlington County College, NJ

John F. Kennedy encouraged Americans to serve their country. To be sure, Burlington County College enabled me to participate in the fulfillment of his dream. My tenure on the Maple Shade Board of Education and my academic career at BCC began the same year.

I knew little about school district operations when I took my seat on the board. Responsibilities, including budget preparation, facility maintenance, and policy review overwhelmed me. My self-confidence declined after each board meeting. Obviously, I had to evaluate my board effectiveness.

My ambition to serve outweighed my desire to retreat. Therefore, I registered for courses at Burlington County College. Oral communications motivated me to accept the board's presidential position. Consequently, my colleagues have elected me to this office for the past six years. Accounting afforded me the necessary skills to decipher financial records. In fact, I served as chairperson of the board's first finance committee. Social sciences provided information on human behavior. Accordingly, I learned to better handle teacher negotiations, policy implementation, and public opinion.

BCC allowed me to understand a school system from a student's point of view. I witnessed the positive impact a dedicated teacher can have on a class. Hence, improving teaching methods through inservice became important to me. I realized the value of extra-curricular activities in developing a well-rounded individual. So, sports and activity expenditures seemed less frivolous to me. I appreciated the importance of state-of-the-art equipment. Therefore, I comprehended that the need for computers and other high-tech machinery justified the purchase prices.

Also, teachers received more compassion from me. They bear the awesome responsibility of educating our country's future leaders. Indeed, when handling student problems, they deserve administrative support. The importance of job satisfaction, fair wages, and work environment influenced my decisions as I balanced labor issues and management issues during contract negotiations.

The academic foundation and positive support from BCC faculty and staff benefited me greatly. Professionally, I transferred to and graduated from Glassboro State College with a bachelor's degree in communications and now work as the college relations specialist at BCC. Personally, I continue to strive for educational excellence as president of our board of education. If not for BCC, my goal to serve my community would not have been realized. Truly, Burlington County College has enabled me to share in Kennedy's dream for a more caring and giving nation.

Linda Dickman graduated from Burlington County College in 1987 with a degree in liberal arts and is now living in Maple Shade, NJ. She is a college relations specialist at Burlington and serves on the board of education in Maple Shade. One of her contributions to her community was to found the Maple Shade Parents' Advisory Council.

Steve Doede
Northcentral Technical College, WI

Male students in the graduating class of 1967, in my hometown of Wausau, Wisconsin, had several alternatives upon graduation: enlist in the armed forces and almost surely be sent to Vietnam; enroll at Northcentral Technical College to pursue a technical education; or, enroll at the branch campus of the University of Wisconsin, both located in Wausau.

Not wanting to be foolhardy, and believing that the technical college was only for those who couldn't make it at a "real" college, I chose to attend the university branch campus.

A year of general studies later, with no particular career objective in mind, I decided that the only thing I liked to do was work on automobiles and that if I was going to pursue that as a career, I might as well enroll at NTC and learn how to do it right.

With the "technical students are second class citizens" thought still in mind, I determined to just attend, learn what I could, and get out of there as soon as possible. Much to my astonishment, I soon learned that the requirements to succeed in technical education were just as rigorous as those at the university. To be successful as a service technician not only did I have to acquire good technical skills, I also had to communicate well, be able to do mathematical calculations, and understand the psychology of human behavior. Surely these couldn't be the same qualities that one would expect from an individual pursuing a profession via the four year university route—but they were.

I also learned that if I were able to do well, there was another profession besides service technician that I could aspire to—that of educator. The thought initially occurred to me as I watched and listened to my instructors, particularly Bob Slane and Wayne Berry. By example, they showed me that instructors can affect students on a level that treats them as individuals, that encourages them to achieve, and that lets them make decisions relative to their own particular circumstances. For instructors to do that requires extreme self-confidence, a positive attitude, and an administration that supports them—something that surely had to exist at NTC.

Inspired by the dedication to excellence of the college, and the professionalism of the instructors, I continued my studies and received a baccalaureate in industrial education two years after graduation from NTC.

Today, as the head of the automotive department of Oklahoma State University's Technical Branch in Okmulgee, OK, with responsibility for 15 instructors and 250 students, I strive to instill those ideals, those philosophies, and those expectations that were first impressed upon me by the administration and faculty of NTC. I am, and will remain, forever indebted.

Steve Doede graduated from Northcentral Technical College in 1970 with a degree in automotive technology and now lives in Okmulgee, OK. As head of the automotive department at Oklahoma State University, he assumes responsibility for supervising 15 instructors and 250 students.

Ben Duran
Merced College, CA

While in high school, my friends and I gave little thought to attending college. Two things happened to change all of that. First, my English teacher continually encouraged us to consider college. Secondly, and most fortunate for me and students like me, Merced College was established in 1962. By the time I graduated from high school in 1966, Merced College was up and running at the local fair grounds with the new campus about to be completed. There were no more excuses that I could use about distance and finances for not attending college.

Since neither of my farmworker parents had graduated from high school and no one in my family had ever attended college, I was more than a bit doubtful about my chances in college. My parents were supportive and encouraging but could offer no financial assistance. In my culture, older sons contribute to the family income while they live at home. Consequently, I took a part-time job that allowed me to meet my obligation to the family while paying my expenses at Merced College. This would not have been possible if Merced College had not been located near my home.

Once enrolling and attending, I was pleasantly surprised to find that quite a few of my friends were also attending. Many came from the same background and had the same kinds of barriers to overcome. In most cases, we were the first from our families to attend college.

Merced College provided me with the confidence base from which I was able to continue my education. After leaving Merced, I completed a bachelor's degree in Spanish and history, followed by a secondary teaching credential. After teaching for five years at my high school alma mater, I completed my master's and earned an administrative credential. The next 10 years took me through a series of administrative and counseling jobs that led to my present position as Superintendent of the LeGrand Union High School District, the very same district that I graduated from in 1966. Currently, my wife and I are both finishing our doctoral studies at the University of Southern California.

My district is 85 percent Hispanic students, but from our graduating class of 1988, over 70 percent are attending college. A majority of those are at Merced College. This is due to our ability to articulate outreach activities with Merced College, providing a place for underrepresented students to attend college. I learned something very special in 1966 about the value of community colleges, and am using what I learned then to ensure that students from my school get a fair shake today.

Ben Duran graduated from Merced College in 1969 with a degree in liberal studies and now lives in Merced, CA. As superintendent/principal of Le Grand Union High School in Merced, he works with administrators, faculty, and students in all areas of the Merced Community College District. In addition to his contributions in working with Mexican-Americans, he is highly recognized by his peers as a capable professional administrator.

Edward M. Eissey
Palm Beach Community College, FL

As a teenager in 1948, I enrolled at Palm Beach Community College on an athletic scholarship. As the son of Lebanese immigrants who settled in Palm Beach County, I strongly felt that the College and the area provided excellent educational and professional opportunities for anyone who worked hard.

Times were tough for my folks and they sacrificed a great deal to help me experience the opportunities that they missed. I was determined to succeed and make them proud of me, and I knew that success started with a higher education.

As a student at a community college, I quickly realized that the faculty cared about the students. We were more than just bodies taking up space in a lecture hall. We all had our own individual needs, and they were addressed by college instructors who were genuinely concerned about our future.

For me, the community college experience allowed for challenge and patience. Instructors had the time to offer one-on-one tutoring when necessary. If a student had a question or difficulty with a subject, there was no problem in scheduling an appointment with the instructor. As time has passed, I still maintain that, while the community college system has grown, it nonetheless allows for a greater margin of individual attention that is not always available at a four-year institution.

I have heard it said that lightning strikes twice, and I am living proof of that concept. In 1978, I became president of Palm Beach Community College, and I still feel as close to the college as I did in 1948. Although 41 years have passed, I am still enthusiastic that the community college system offers the best possible education at affordable prices. In some instances, a community college is economically the only choice some students can make. However, while the college may be less financially taxing, it is nonetheless an institution of higher learning that mandates the highest levels of academic achievement for the successful completion of any degree or certificate program.

I am very proud of the education I received at Palm Beach Community College and I am equally as proud to serve as this great institution's president. Pride is what makes a student a scholar, a painter an artist, or a teacher an educator. I am proud of PBCC and I am proud that my children followed in my footsteps and attended the college as well. I look forward to seeing my grandchildren at this college as they pursue their higher education.

The community college experience ... it is without a doubt one of the best choices anyone could make when planning their educational future.

Edward M. Eissey graduated from Palm Beach Community College in 1948 with an associate's degree. He is currently president of Palm Beach Community College and lives in North Palm Beach, FL. He is director of St. Judes Hospital for Children in Memphis and serves on the board of directors of a Palm Beach bank. He is also a recipient of the City of Hope Humanities Award and a North Palm Beach Chamber of Commerce Appreciation Award.

Geraldine A. Evans
Rochester Community College, MN

My family lived on a small dairy farm in Southeastern Minnesota. Although there was never much money, there was always an interest in reading and the larger world around us. Several very admired relatives were teachers, and I never remembered a time when I did not plan to attend college. Thinking back, I sometimes wonder why I had such faith in that reality when the cost was really quite beyond our means.

I graduated from Rochester Senior High School with a good academic record. The local community college was housed in the fourth and fifth floors of the high school building, so it seemed very natural to "move upstairs to the JC" when the decision came to choose a college. Many of my friends also attended the junior college, and my busy activity schedule and rigid academic pursuits just continued in a very spontaneous manner.

Although I know I thought I was very mature at the time I graduated from high school, I often look back and think of how naive I really was, and how privileged I was that my community had established such a fine college. My effortless decision to attend Rochester Community College was a very fortunate one for me. It allowed me to live at home and obtain a first-rate college education at a very low cost. I was able to work 20 to 25 hours a week to pay for my tuition and books, as well as set aside some money for the upcoming two years at the university.

The superb faculty of the college was intellectually stimulating and the quality and reputation of the courses allowed me to actually gain credits on transfer to the University of Minnesota. I can truthfully say that the best teaching I ever received was provided at RCC. The classroom learning, however, was the second most valuable aspect of my RCC experience. As editor of the college newspaper, as a member of the student government, and as an officer and member of several other clubs and organizations, I learned the skills of leadership, teamwork, group participation, and cooperation toward common goals that have been invaluable in progressing in my career.

The same sort of natural progression of events led me back here to be president of Rochester Community College. It is my sincere hope each day that this college is having the same very positive impact on the lives of students today as it did when I was attending.

Geraldine A. Evans graduated from Rochester Community College in 1958 with an AA degree and is now living in Rochester, MN. She is the president of Rochester Community College. Dr. Evans has been very involved in Rochester's effort to form a university center and is trying to increase continuing education offerings, student services, business outreach, and programs for disadvantaged students.

Lacinda Drake Evans
Genesee Community College, NY

After sustaining a work-related injury a number of years ago, I found myself disabled. For two years I searched for work that my body could endure, but I was told by many prospective employers, "I'm sorry, but you don't have the education required for this position." After much frustration and damage to my self-esteem, I was referred to vocational rehabilitation. With the aid of that agency, I applied to Genesee Community College.

The steps I took as I entered the doors of Genesee Community College were the most frightening of my life. I had been out of school for many years and had no idea of what direction to take. However, the advisement staff at GCC were warm and supportive. Working on an individual basis, they helped me make out my first semester's schedule and a plan for other classes in later terms. Although I had arrived terrified and full of doubt, I left with some encouragement and hope.

I started my first semester with virtually no self-confidence. My home environment provided no support or encouragement. The transition from mother and housewife to mother, housewife, and student had totally overwhelmed me.

I don't think it would have surprised me if I had failed. Because of the nurturing atmosphere of Genesee Community College, however, I made tremendous growth that first semester. I gained confidence in my ability to succeed, and I was able to set goals in a direction that at one time had seemed completely unreachable. I had regained control of my life.

Genesee Community College allowed me to gain strength and heal the wounds of my past. It helped me release an inner motivation that I never knew existed. It helped me see that life is full of choices and that, although I will always face some obstacles, I now have the ability to overcome them. I am extremely proud to say that I am a graduate of Genesee Community College—a place that made my dreams come true!

Lacinda Drake Evans graduated from Genesee Community College in 1990 with an associate's degree. She now lives in Perry, NY, where she teaches arts and crafts in that community's summer youth program. She also helped initiate an arts and crafts program for mentally retarded adults and is preparing a crafts program for senior citizens.

Michael H. Farmer
Greenville Technical College, SC

I am one of those fortunate persons who make their livelihood enjoying themselves. As a high school chemistry teacher, I experience challenges and rewards every day, find fulfillment and satisfaction, and realize goals and accomplishments. I feel that my job is important—a link between past insights and future discoveries, youth and maturity, problems and solutions. My work is both career and passion—which after 17 years shows no sign of diminishing.

The path to contentment was not direct, smooth, or easy. Thirty years ago, as a misdirected high school graduate, I had no intention of continuing my formal education. My parents and peers saw no value in more schooling and believed that getting a job was the best choice for me. I generally agreed, applied to work at a local auto battery factory, and resolved to "work hard and make good."

It was a supervisor who interrupted my unambitious plans, encouraged me to enroll in the chemical technology program at Greenville Technical College, and set in motion a series of events that were to change the direction of my life. The first course in chemistry led to others; part-time classes became a full-time program; random study evolved into degree commitment. I was elected student body president and served as editor of the first yearbook. I set new goals: more technical expertise, a better paying job, improved social status. I graduated with honors from the associate degree program, became a supervisor with a synthetic fibers company, and settled back to reap the benefits of my accomplishments.

My "settling" was short-lived. The classes at Tech proved to be a beginning rather than an end; the instruction I received was a catalyst rather than conclusion. My instructors encouraged me to seek higher goals, and when Clemson University opened a two-year extension program on the Tech campus, I enrolled in a baccalaureate degree program. When I received my degree in chemistry two years later, I again sensed the event was not a conclusion, but another beginning. During my years at Tech and Clemson, I became convinced of the importance of education and felt a desire to become a part of the profession that uniquely touches the future. I chose to become a teacher.

Michael H. Farmer graduated from Greenville Technical College in 1970 with a degree in chemical technology and is now living in Tigerville, SC. He is a high school chemistry teacher at Riverside High School in Greer, SC. He received the Human Relations Award from the Greenville County Education Association, the Award for Excellence in Secondary Science Teaching from the South Carolina Academy of Science, and the Presidential Award for Excellence in Science and Mathematics Teaching.

In the 17 years since I made that decision, there have been no regrets, only reassurances. I've completed a master's degree program, begun work on a doctorate, received numerous state and national awards, and influenced hundreds of young people. I am forever reminded of how important a teacher's role is in influencing others. I hope that I can perpetuate what my Tech instructors taught me—that our possibilities are limited only by our degree of determination.

Mark J. Felsheim
Madison Area Technical College, WI

A commitment to learning. A commitment to teaching. A commitment to education. These are focal points of my life and my career so far. And, when I ask myself where this commitment came from, I have to answer honestly that much of it came from my experiences at Madison Area Technical College.

I have always enjoyed reading and learning and when I was younger I had also enjoyed school. Somewhere along the line, however, I lost that feeling of joy. By the end of high school, school was just a place to meet friends and to pass the time. I drifted for a year after my graduation with the vague notion that I should go on to college, because that's what you were supposed to do, but with no real desire to do so.

I started my studies at MATC for several reasons. First, the liberal studies major gave me a chance to pick up basic courses while I waited for the vision to tell me what to do with my life. Also, their flexible schedules allowed me to continue to work full-time or nearly full-time while I attended school. What I discovered at MATC was that learning could be fun again.

With the personal attention I received and the enthusiasm of my instructors, I was able to raise my "B" average grades in high school to honor grades at MATC. I found the experience stimulating and challenging and fun.

When I later transferred to the University of Wisconsin, I found I not only could compete with students who had already had two years of a university, but that I could surpass many of them. Much to the chagrin of friends who had dismissed MATC as "Tinker Toy Tech," I was also an honors student at the university.

Since my graduation I have taught high school, junior high, adult basic skills, adult high school, and associate degree college level. I have been a curriculum specialist, an inservice director, a university extension teacher, a business consultant, and an administrator, but I have never forgotten my experiences at MATC.

I still feel a strong link with MATC and I am still grateful for the education and the experience. I am a member of the alumni association executive board and I have served on advisory committees for the college. In my current position with Southwest Wisconsin Technical College I have been able to draw on my experiences at MATC and my friends there to make this job a more rewarding experience.

> *Mark J. Felsheim graduated from Madison Area Technical College in 1975 with a degree in liberal arts and now lives in Fenninore, WI. He is dean of outreach and educational services at Southwest Wisconsin Technical College. He is a member of the Wisconsin Vocational, Technical and Adult Education Instructional Administrators and Continuing Administrators Association and serves on the Executive Board of Madison Area Technical College.*

James Ford
Lower Columbia College, WA

Community colleges are many things to many people. A myriad of successful individuals have gotten their starts, the stimulus they needed, the educational tools needed, the necessary encouragement, and the opportunity to fulfill their dreams at community colleges. In my case, all of these factors were important, plus the opportunity to develop socially and to participate in college activities.

I realized that I would need to study in a college where classes were small and faculty would provide individual attention. I wanted a college where one could participate and excel in athletics, work on the college newspaper and annual, and really get acquainted with friends and faculty. In summary, I wanted a smaller college that had excellent faculty who would work closely with me, a college with a friendly atmosphere where close friendships could be formed, and a college that cared about me as an individual.

All of this I found at Lower Columbia Junior College (now Lower Columbia Community College) in Longview, Washington. In 1947 I enrolled at LCJC under the GI Bill and lived at home. LCJC did not have a campus at that time, and classes were held at various sites around the city. Headquarters were in the basement of the public library. But a well-defined campus was not necessary because of the type of school spirit and the caring educators who taught at this college.

It was there that I received outstanding instruction and encouragement from a dedicated and well-credentialed faculty. For example, LCJC had two of the finest biology instructors I have encountered in my entire career. They held doctorates in zoology and provided the encouragement and self-confidence I needed to proceed in college and eventually to earn my doctorate in cell biology in 1962.

I can't thank this college enough for what it did for me and my career. The faculty, coaches, administrative staff (president, registrar, and a dean at that time), and fellow students all cared about each other and provided an environment and a spirit where success was encouraged.

And it was here, at LCJC, that I actually made my career choice. It was at this college where I had such a pleasant and rewarding experience that I determined that I wanted to work in a community college. I didn't realize then that I would become a community college president and work to provide the same environment for the next generation. But this did happen, and I would not change it for anything.

Today, LCCC has a very fine, well-equipped campus in a beautiful part of Longview, but the atmosphere is still the same—a friendly, inviting, excellent, and caring faculty and staff providing opportunities for students of all ages.

James Ford graduated from Lower Columbia College in 1948 with a degree in biology and now lives in Mount Vernon, WA. He is president of Skagit Valley Community College, where he has produced many professional publications. Among his academic contributions has been serving as senior author for the textbook, Living Systems: Principles and Relationships.

Gary Fouraker
Southeast Community College-Milford Campus, NE

Having spent my childhood in a lower-middle class farm community with no other family members who attended postsecondary school, the thought of attending a traditional four-year college was the furthest thing from my mind. While I received excellent grades throughout high school in a college prep curriculum, my interests were more technical, primarily aimed at accounting. After considering several area business schools, I decided that SCC-Milford offered the best educational opportunities in my area of interest, along with the additional curriculum in automated data processing.

It was during my second year at SCC that one of my instructors convinced me that I should pursue additional education in order to enhance my career possibilities. He was instrumental in my being hired as a part-time computer programmer at the University of Nebraska while enrolling in the college of business administration. That was only the beginning, since I continued to work full-time as a systems analyst and in other managerial positions at the university while attending classes part-time, eventually earning bachelor's and master's degrees.

My education and experience at SCC were invaluable, providing me with the resources to continue my education, and allowing me to use my knowledge and expertise in the computer and data processing areas to automate a variety of systems and procedures throughout the university. Many of the ideas and techniques developed during this time have been shared with colleagues through presentations and papers at various national forums, and have hopefully resulted in improvements at other institutions.

While computer equipment technology has changed dramatically, the basics of computerized data processing, learned almost 25 years ago at SCC, still serve me well today as I attempt to manage the business affairs of a major intercollegiate athletic program.

> *Gary Fouraker graduated from Southeast Community College-Milford Campus in 1966 with a degree in electronic data processing and now lives in Lincoln, NE. He is assistant athletic director for business affairs at the University of Nebraska. Gary has been active in various civic organizations, including the Village of Hickman Planning Commission and the College Athletic Business Managers Association. He has also selected and integrated computer hardware/software into functions of his department.*

Martha Foxall
Richland College, TX

My life changed when I received a phone call from my son's math teacher at Berkner High School asking if I wanted to continue my education. My answer was "Yes, I do." I was the mother of five children, rearing a four-year-old daughter at the time.

I did not know what I wanted to do. Just starting all over again felt good.

I did not have the money to pay for my courses, so I was given a grant. I was also not sure if I could go to school on campus, because of my four-year-old daughter, but the college had an extension in my community. I put my daughter in a day care center and enrolled.

I was not sure about myself. I began a reading course. My instructor was Carl Denmon. He encouraged me to keep going.

I wanted to stop because I felt alone. I was the oldest among the group, and it was not easy to sit there each day, making mistakes. But the students who were in the class said, "Do not stop."

Later, I started a writing course. Writing was a course I thought would be easy, but it wasn't. My instructor, Jane Peterson, worked very patiently with me. She would work early or late to help me get my subject completed. As it turns out, I still need writing courses.

During my college career, I was informed about the degree plan. A student could take 60 hours and get her/his degree in whatever field she chose to work in. All the courses I needed to become an educational paraprofessional were taught at the college.

The most rewarding experience for me was that my instructors, including the ones in the paraprofessional program, never gave up on me. They kept me informed about what I needed to study.

I did graduate with an associate degree in applied arts and sciences in 1980.

In 1979 I was hired by the Dallas Independent School District as a teacher's aide. Now I work with the Thomas Jefferson High School in an Alternative Education Program. I work with students who have behavior problems.

Sometimes there is a need to sit and listen to students tell their side of the problems. I do this and give instruction. I also have clerical responsibilities.

Thanks to the Alumni Association for giving me an opportunity to share my story.

Martha Foxall graduated from Richland College in 1980 with an associate's degree in education and now lives in Dallas, TX. She is a desegregation aide at Thomas Jefferson High School and began classes in Hamilton Park—Richland's first "storefront" classroom. She is also active in her precinct and community.

Randy R. Griffin
Treasure Valley Community College, OR

With encouragement from my parents to pursue a college education, I found it challenging and exciting deciding which educational institution I wanted to attend. However, coming from a very small rural community, I also quickly found this process to be intimidating to say the least. Therefore, I chose my local community college, Treasure Valley Community College, because of its small size, good reputation, availability, and because it more than met my needs.

I soon learned that this decision was a sound one for several reasons. The ratio of students per faculty member was low in contrast to the university setting. For me, this proved invaluable. I quickly found all faculty and staff to be humanitarians ready to assist all students in a caring and dedicated way. All would go the extra mile if necessary either in academics or personal endeavors. The personalized attention I received during my two years at TVCC was unsurpassed. The support, enthusiam, and positive atmosphere quickly rubbed off on me, and I gained confidence from day one. Once this momentum began to build, it was like a giant steam roller, which was virtually unstoppable. I was encouraged, stimulated, and felt as if I was on top of the world.

It was easy to become involved in campus activities where faculty and staff took part like one big family. This promoted a unified and caring feeling within the college that flowed directly into the community and into my personal life.

Looking back, I am now fully aware of the dramatic impact TVCC had on my life. Upon graduation in 1975, with an AS degree in business administration, I had in two years overcome my self-esteem problems and was able to meet all future challenges, both educational and personal, with enthusiam and confidence. It provided me an excellent and necessary stepping stone to successfully complete a BS degree in business/economics in 1977 as well as convincing me that education is truly a life-long learning commitment.

In 1984, I had the fortunate opportunity to join the administrative staff of TVCC. Filled with nothing but positive memories about TVCC, I actually felt again that I was the luckiest person in the world. Now after six years in this position, I still feel this way and have an even greater sense of accomplishment because I help provide educational opportunities to individuals of all age groups and am able to contribute heavily to my community.

TVCC is still a great community college providing the same rewards I was lucky enough to obtain 14 years ago. The day I decided to attend TVCC has proven to be one of the best decisions of my life.

Randy R. Griffin graduated from Treasure Valley Community College in 1975 with a degree in business administration. He now lives in Ontario, OR, and serves as the business manager of his alma mater. In the community, he works with the local Lions Club in various projects, including drug awareness and prevention programs and eyesight programs. He is also very active with the TVCC Foundation to provide scholarships for students.

William Alexander Hall
Tri-Cities State Technical Institute, TN

Ifeel as if I have been in education all my life, both as a student and as an instructor. I was among the first group of students who enrolled in what was then the tool design drafting program at Tri-Cities State Area Vocational-Technical School (now Tri-Cities State Technical Institute) in 1966. Now, over 20 years later, I am on the other side, as an assistant professor at the same school.

In 1966, I was looking for a skill at which I could make a living and one that I would enjoy the rest of my adult life. My background was rich in farming, due to the fact that my ancestors had been farmers in Northeast Tennessee for five generations. I had a need for a change, one that was challenging, but still allowed me to be creative. My goal did not include leaving the region to find fame and fortune. To me, my fortune was right here with my family and friends.

My experiences at Tech taught me much more than the essentials of the drafting trade. I learned that pride in workmanship is not out-of-date; that an honest day's work is as important in the business world as it is on the farm. I learned that professionalism is the benchmark of any job.

My time as a student at Tech gave me tremendous respect for the teaching profession. The intimate atmosphere of a small school gave me the chance to work closely with my classmates and teachers. These instructors motivated me greatly. My instructors at Tech were more than teachers. They were role models, advisors, counselors, and friends. I often think of them now as I deal with my students. I hope that as a teacher, I am able to instill some of these values in my students.

I only recently discovered that one of my ancestors was Samuel Doak, founder of the first school in Tennessee. I'm sure he must have loved teaching as much as I do.

William Alexander Hall graduated from Tri-Cities State Technical Institute in 1966 with a degree in tool design drafting and now lives in Blountville, TN. He is an assistant professor at Tri-Cities State. One of his ancestors is Samuel Doak, founder of the first school in the state of Tennessee.

Barbara J. Hamilton
Central Piedmont Community College, NC

I had my first child at age 13, so I had to drop out of school. Having a child at that age and in those times was very hard. People look down on you and don't respect you. Although I did get married after my child was born, it was not easy. I didn't want to go back to school, so I started to look for a job. Without any training, the only jobs I could get during those days were factory or domestic. I chose domestic.

In 1960, I moved to New York. Again, I found that it was hard to get a decent job without any training, but you could still find work. I lived there for 22 years and had jobs as a domestic, selling books, inspecting zippers, in banks, as a receptionist and phone operator, in department stores; my last job was with the New York State Department of Taxation Finance. While working at the bank, one of the requirements for keeping the job was to take a class to prepare for my high school equivalency. After I received that, it was easier to get better paying jobs.

After living and working in New York for those years, I decided to return to North Carolina to raise my youngest child, who is now 12 years old. I chose Charlotte because it offered more job opportunities than Durham, my home town. Again, I faced the same problem of not being able to get a decent job.

I realized, after working at a variety of jobs, that I wasn't going to get anywhere unless I could type, use a computer, or get some kind of a degree. After checking around, I found out about a typing class that was being given at Double Oaks Center, one of Central Piedmont Community College's off campus locations. This was my first step toward my present position. I didn't do too well in the typing class, but it got me interested in trying to better myself, and that is what CPCC is all about. CPCC has all kinds of classes and seminars to encourage people to come back to school. Their Human Resources class was such a class. It shows you what you have to face in the work world today and helps you prepare yourself for this by teaching you how to dress, how to prepare a resume, and how to get a job interview. All of these factors combined gave me the incentive to try to continue my education.

During the second quarter, I didn't know what courses to take or what kind of degree I wanted. After talking to other students, I found that CPCC had counselors who would help you. The counseling office assigned a counselor to me and with her help and suggestions I set out to get a degree in general office technology. The leading subject was typing. I stayed with it because I found that the instructors and everyone at CPCC cared about the students.

With my degree in general office technology, which I received in May of 1989 at age 50, I received a position with CPCC. I would like to thank everyone at CPCC for all of their help and encouragement. It has been a great experience!

> *Barbara J. Hamilton graduated from Central Piedmont Community College in 1989 with a degree in general office technology and now lives in Charlotte, NC. She is a secretary at Central Piedmont and is known widely for her helpful service-oriented approach to the college's students and employees.*

Minh-Anh Hodge
Columbia Basin College, WA

I was an interpreter for the United States Air Force in Vietnam before I came to the Tri-Cities with my husband in 1973. After the fall of South Vietnam in 1975, I began working for the Kennewick School District as a consultant for Southeast Asian refugees, and later taught classes in the area of English as a Second Language. Because of my desire to make decisions that would positively affect students, I decided to go back to school to earn my teaching degree.

In 1981, I enrolled at Columbia Basin College and Eastern Washington University simultaneously. Even though I was a homemaker with three small children and had a full-time job, I graduated from CBC in 1983 with a 4.0 GPA. I obtained my bachelor's degree in education in 1984, graduating in the top 20 from Eastern, and received the Mary Shields Wilson Medal for Academic Excellence. My master's and principal's credentials were obtained the following year.

For the past five years, I have been coordinating the Kennewick School District bilingual programs and I truly enjoy working with the students, staff, and parents to provide a quality educational program for all students, especially those who are limited in speaking English.

I am going back to college for more advanced studies. No matter what kind of academic achievements I reach in the future, I will maintain a deep appreciation for the excellent education I received from the caring staff at CBC. They provided me with the solid basics and support that helped enhance my future learning and opportunities.

Minh-Anh Hodge graduated from Columbia Basin College in 1983 with an AA degree and now lives in Pasco, WA. She is director of the bilingual program of the Kennewick School District and is heavily involved in advancing educational opportunities for youth in the Tri-Cities, especially those with limited English backgrounds. She also chairs the board of trustees of Columbia Basin College.

Cynthia Ingram
Patrick Henry Community College, VA

Patrick Henry Community College has provided the opportunity for me to continue my education and experience in tremendous career challenges. My initial experience coming to the college was that Patrick Henry Community College was a warm and friendly place to learn. My feeling about the college has not changed one bit. Two of my favorite instructors, Whitey Pitts and Martha Prillaman, took special interest in assisting, advising, and encouraging me to excel in my classes.

When I decided to return to college, I was 32 years old, married, a mother of three, and a full-time shift worker at Dupont Company. I was really very indecisive about my future. It had been more than 10 years since I had been in college. I am convinced that the support I received at PHCC increased my confidence in my own knowledge, skills, and abilities, which has led me to really believe that I am only limited by my own imagination and sacrifices.

Cynthia Ingram graduated from Patrick Henry Community College in 1981 with a degree in business administration. She acts as coordinator of extended learning services at Patrick Henry Community College and now lives in Martinsville, VA. Under her leadership, the college has developed extensive training programs for local employers and established a center for small business. She also serves on the board of directors and executive board of the local Chamber of Commerce and the Community Employment Advisory Council.

Margaret Ingram
Indian River Community College, FL

My greatest contribution to society has been primarily in the field of education. My dedication to this challenging profession has been kindled by the satisfaction I derived from continuing my own education despite a most humble beginning. It was my community college experience that gave me the boost of confidence and intellectual aspirations to fulfill my goals. The year I received my associate's degree, 1976, was a major turning point in my life, and my memory of the occasion is as vivid as yesterday.

Let's for a moment revisit an incident on campus early in my educational journey. It was after my two sons were born that I began my studies at Indian River Community College with the support of my husband, a business management student. Shyness prevented me from seeking financial aid, but my husband had made an appointment with the dean to discuss our financial situation. Out of concern for my extreme shyness, my husband answered all of the questions at this meeting including those directed to me, prompting the dean to inquire, "Mrs. Ingram, can you speak?"

Two years later I received my associate's degree and two years after that, my bachelor's degree in elementary education. It was the personal interest and encouragement I received at Indian River that enabled me to overcome my shyness and lack of confidence. The intimate environment of the community college made me secure in my abilities, as did the feeling that everyone was there to help and support my endeavors.

I found that obstacles that seem like gigantic mountains are chiseled into sand when you become part of the community college family. Since that humble beginning, I have never again underestimated my ability to achieve any goal in life.

For example, whoever would have dreamed that just 12 years later, this shy and quiet girl would become 1988 Florida Teacher of the Year and Christa McAuliffe Ambassador for Education? Who would have thought I would be chosen to fill a position that allowed me the opportunity to travel around Florida to promote education, to spread my zeal across the state and nation to "Reach for the Stars Through Education?"

One year later, this community college graduate was featured among the nation's finest in the "Disney Channel Salute to the American Teacher," and the same year, served as a participant in the Governor's Education Summit to help determine the direction of education in the State of Florida.

What a wonderful opportunity to say thank you to a paramount institution that made my humble beginning a great beginning, my community college experience.

Margaret Ingram graduated from Indian River Community College in 1976 with a degree in elementary education and now lives in Vero Beach, FL. She is a 4th and 5th grade math and science teacher at Beachland Elementary School. For the past ten years she has served on various school advisory and curriculum committees, the district's Peer Teacher Program and Parents' Involvement in Education Tutorial Program, and sponsors the Literary Club and Values Committee at her school. She was selected as teacher of the year by two organizations in 1986 and 1988 and also belongs to numerous civic organizations.

Christine Johnson
Fresno City College, CA

When I arrived at Fresno City College in the Fall of 1963, the idea of attending a community college was still foreign and hard for me to accept. I had just graduated from high school in a rural Texas community, with great expectations and the long-held belief that I would attend one of the historically Black colleges along with my friends.

During the summer, however, all of my plans changed. Due to my father's illness, instead of going to a major four-year college, I was stuck at what I believed to be a second-rate junior college in a place I was convinced no one had ever heard of.

While I understood that the family's income could not augment my scholarship, and everyone had to make sacrifices for the next several years, I still felt cheated. Nevertheless, the next two years at Fresno City College literally transformed and helped to shape my attitude toward life.

For example, when I arrived at Fresno City College I knew I wanted to go to college, but I was not completely focused on which career I wished to pursue. Reality soon set in, however, and I felt compelled to focus. I decided that due to the family's financial picture, I should concentrate on a career that would get me out of school in four years and into the labor market. Somehow, I was directed toward the counseling center and my transformation began.

My experience with the counselors at Fresno City College helped me to learn the art of setting goals and persisting until those goals are achieved. In the process I developed, internally, the motivation and determination to succeed in all aspects of my life.

My experiences in the classroom were equally motivating. The no-nonsense instructors provided clear and concise expectations about my performance. I was challenged to strive for excellence. My mind expanded by leaps and bounds as I was introduced to the different subjects in the liberal arts curriculum at Fresno City College.

My positive experiences at Fresno City College from 1963-1965 enabled me to attain an associate degree in liberal arts and make an easy transition from a junior college to a four-year institution such as Fresno State. At Fresno State, I earned a bachelor's in social welfare and a minor in psychology. After a few years in the work force, I returned to Fresno State and earned a master's degree in counseling and guidance. However, the quest for knowledge implanted and nurtured at Fresno City College continued to grow, and I felt motivated to return to school at the University of Southern California to pursue a doctorate in higher education administration.

The solid foundation I gained at Fresno City College has sustained me professionally and personally through the years in my career as a counselor, educator, entrepreneur, college dean, and consultant.

Christine Johnson graduated from Fresno City College in 1965 with a degree in liberal arts and now lives in Fresno, CA. She is an administrator at Fresno City College and belongs to numerous professional organizations. She has given many professional presentations and served diverse organizations in the San Francisco Bay Area for the past dozen years as a grant writer and board member.

Lynn Johnson
Clark State Community College, OH

As my high school years came to an end, I was unsure of what my future held. My parents suggested I try Clark State Community College so I could continue to live at home, work, and get an education.

I began my first year at Clark State taking general classes part-time, planning to go into social work. After getting through my first human services class in my second year, I decided social work wasn't the place for me. The second quarter of that year I met a wonderful advisor who started me on my way to a degree in early childhood education.

For the next year and a half I was unsure if I could make it. There were so many projects to do and so much information to learn. But my advisor kept encouraging me and I became determined to graduate.

I began my first practicum in a wonderful preschool program and started a new job working in an infant-toddler program designed to care for the young children of teenage parents who were finishing high school. In addition, I was waitressing on the weekends. I soon became overloaded by the two jobs, the work on my practicum, and my classes — but I made it through.

Finally I had made it to my last quarter at Clark State. This meant I was to start another practicum at a different center. The thought of leaving the previous preschool program broke my heart. I went to my advisor begging and pleading to stay. She said it was policy for each student to switch sites for the second practicum. She finally decided to compromise and said if I wrote a proposal on the advantages of remaining there I could stay for my second practicum. I wrote the proposal and spent my final quarter with the same program.

Before I knew it, graduation came. I took a summer job, unsure of what the future held. My supervisor at the infant-toddler center approached me at mid-summer with a proposition. She was resigning and wanted me to replace her. I wasn't sure if a new community college graduate was ready for a day care director's position. I thought long and hard about the offer and decided to take a chance. I interviewed for the job and was hired the next day.

I could not have taken on this challenge without the help of Clark State Community College and my advisor. Because of them I am using my education to keep teenaged parents in school and help shape the lives of small children whose only hope may be the care I provide.

Lynn Johnson graduated from Clark State Community College in 1989 with a degree in early childhood education and now lives in Springfield, OH. She is now director of a day care center, thereby taking on the challenging responsibility of maintaining the well-being of infants and toddlers who often live in less than ideal home situations.

Patsy Joyner
Paul D. Camp Community College, VA

When Paul D. Camp Community College opened in 1971, I had been out of high school and married for 10 years. I was also working full-time as a medical secretary and was actively involved in a number of community organizations. Frankly, I did not want to give up any one of these roles. However, I did want to pursue my education because my experience had shown me that having credentials is the best way to make a difference in your own life and the lives of others around you. I wanted to have an opportunity to make a difference.

My first contact with the community college was my meeting with a counselor. She was warm, friendly, and encouraging. I signed up for a full load and attended classes four nights a week and during lunch hours. With this hectic schedule I learned to juggle many things to make each minute count. I ate lunch in the car, read notes and polished my nails at stoplights, and studied after my husband and daughter went to bed. I remember one particular time when I was studying for exams. I was so intensely involved in my psychology notes that I lost track of time. Suddenly I heard a bird chirping. I went to the window and raised the shade. To my surprise it was the next day, and I was due at work in one hour! I was so motivated by my experience at the college that I happily went without sleep.

At PDCCC I was challenged to become all I was capable of becoming. Each of my professors and advisors took personal interest in me. They bolstered my confidence and self-esteem. Because of their support I was able to graduate with the first graduating class at PDCCC (with honors). It was not an ending, but a beginning. I left with a thirst for more!

I went on to earn bachelor's and master's degrees from Old Dominion University and the educational specialist certificate and doctorate of education degree from the College of William and Mary. I am especially proud that I accomplished this while working full-time and being actively involved in community and professional affairs.

Today I am where I belong and where I can contribute the most: I am working at Paul D. Camp Community College! I feel fortunate to have the opportunity to provide the encouragement that I received to others. I especially like being a role model for other women. I want them to know that they can be a wife, mother, community contributor, and a student, and that the best place to begin is the community college!

Patsy Joyner graduated from Paul D. Camp Community College in 1977 and is now living in Courtland, VA. She is the director of community and continuing education and public information officer at Paul D. Camp Community College. She is treasurer of the State Executive Commission for the Virginia Identification Program for Women in Higher Education Administration, member of a statewide Adult Literacy Commission, Virginia representative to the District II Board of Directors for the National Council for Marketing and Public Relations, and chair of the Deerfield Correctional Center Advisory Board.

John Long
Technical College of the Lowcountry, SC

Attending college had been one of my life-long goals. After completing high school, working for a couple of years, and spending better than two years in the military, I was more than ready to go to college. I had discovered that in order to get the type of job I wanted, I had to have a college education.

Finding the right college, however, was a different story. The type of college that I had always envisioned was one where the classes were small and the instructors were caring, a place where I would not become just another social security number. I also had to be able to work full-time to support my family.

I found everything I needed at the Technical College of the Lowcountry, a two-year technical college in Beaufort, SC. I found a job in the maintenance department of the college, which allowed me to attend classes full-time. Once I started classes, I was motivated and encouraged at every turn, not only by the instructors but also by everyone I came in contact with at the college. The type of atmosphere I found at TCL was a deciding factor in my plan to continue my education after receiving an associate's degree in 1974.

After I received a bachelor's degree in business administration in 1976, I applied for a position as a developmental mathematics instructor at the college. I was hired. Since then I have seen a lot of students who are mirror images of myself when I first began my pursuit of a college education. These students lack confidence, motivation, and a belief that they can accomplish their goals. I try very hard to give to them the same encouragement and special attention that I received.

I am currently enrolled in a graduate degree program. Without the encouragement and special treatment I received in a two-year college, I probably would not have the confidence or motivation to successfully complete the graduate program. My experience in the two-year college system has certainly helped me to better serve the students I teach and advise.

John Long graduated from Technical College of the Lowcountry in 1974 with a degree in business and now lives in Beaufort, SC. He is a developmental education instructor at Technical College of the Lowcountry and has provided support and encouragement to many students entering developmental mathematics.

Lori A. Manion
Scottsdale Community College, AZ

Scottsdale Community College was a natural step toward a well-rounded college education. As a recent high school graduate, I was intimidated not only by university life but by overcrowded classes and confusing curriculum choices. I decided to enroll in a local junior college to help me adjust to college life, while keeping expenses to a minimum.

The instructors and staff at SCC offered me a variety of opportunities to become an active member of the college community. I was impressed with the availability of smaller classes and personalized curriculum advisement. While attending SCC, I was able to investigate many academic programs. After enrolling in an introductory geology course, I realized that I had found a curriculum that was both interesting and exciting. I had taken the majority of available geology courses while attending SCC, and I had been actively involved in tutoring and the geology club. I was enjoying the earth sciences for the first time, and I was eager to continue my education at the university level. Upon graduation from SCC, and with the support of my geology professor, academic advisor, parents, and friends, I decided to set new goals for my education.

I confidently enrolled in the mountain campus of Northern Arizona University and with perseverance earned a bachelor's degree in earth science education. The junior college experience was not only a natural educational progression for me, but also an enjoyable experience. It gave me a chance to grow and mature both mentally and emotionally while sharpening necessary skills for college survival. My SCC geology professor and academic advisor were the two most influential role models for my career selection. I shall always be grateful for their valuable assistance and encouragement.

Lori A. Manion graduated from Scottsdale Community College in 1984 with a degree in science and now lives in Scottsdale, AZ. She is a teacher and a volunteer for the Liberty Wildlife Rehabilitation Center. She received the Outstanding Geology Student Award in 1984 and served as a student representative to the Honors Program Commission.

Betty McKie
Morgan Community College, CO

As community colleges came of age in the 1970s, so did I. And I think I faced many of the same growing problems that community colleges encountered. As a 17-year-old high school graduate, I went off to the university down the road—not because I knew what I wanted to do, but because all my friends were going there. Nobody wanted to stay at home and attend the newly opened community college.

I found myself drifting along that first year, homesick and not sure of where I should be going. After a year I gave it up and came home, found a job (for a little more than minimum wage), and settled down. All good things eventually come to an end and so did my job. After two-and-one-half years the company folded and moved out of town. I was lucky to have gotten that job and it wasn't long before I decided I would have to acquire some new skills if I was going to get another. The local community college now seemed like the place to go.

I walked into registration at Morgan Community College with the idea that I would take a shorthand class that would help me in the job market. An hour later, after sitting down with an advisor, I enrolled not only in a shorthand class but in four other classes as well. I was a full-time student and my real college experience had begun.

Somewhere at Morgan Community College the quest-for-knowledge bug bit me and I couldn't get enough. I studied hard, participated in student government and Phi Beta Lambda, served on committees, and got involved in everything possible. I had truly found a place where I could be all that I wanted to be. MCC and I grew together and we haven't stopped.

Ten years later, having finished two associate's degrees, a bachelor's degree, and a master's degree, I am back at MCC—this time as a faculty member and chair of a division. It is my turn to help students discover themselves and their successes by sharing my experiences over the years and supporting them in their endeavors.

I believe it is true that we are a composite of our past experiences. I carry with me a part of all the wonderful teachers I had, the staff at the college, and even the president, who always took the time to stop and talk about the future with me. I am what I am today because Morgan Community College created an atmosphere for me and for many others to thrive in.

Betty McKie graduated from Morgan Community College and earned degrees in business in 1978 and 1979. She now lives in Fort Morgan, CO. She is chairperson of business and technical education at Morgan Community College. She is a state Phi Beta Lambda officer team advisor, works on numerous state committees for business education, chairs the core curriculum writing team for the community college system in Colorado, and was named Faculty Person of the Year at MCC in 1987.

Maria Lopez Miller
McLennan Community College, TX

Education was very important in my growing years. I grew up in a bilingual home where Spanish and English were the languages spoken. My mother spoke very little English, so it was out of necessity that I, being the oldest of four children, learned as much as possible so I could help at home. My father provided as much educational material as his salary could afford.

Since there was not much money in our household for toys, my sisters and brother and I spent much time playing such games as school and password. This contributed a great deal to my desire to teach. As far back as I can remember, I wanted to be a teacher. Having the support of my parents, I did very well in school and was given a scholarship to attend McLennan Community College. My dream was coming true.

I felt both nervous and excited about attending college. My fears were relieved as I attended my classes and met my instructors. They provided genuine warmth and interest in helping me grow. With my grades and marvelous help from the financial aid director at MCC, I had the good fortune of receiving a Ford Foundation Scholarship. This scholarship helped pay for my next two years at Baylor University.

I have been teaching for 16 years now and have since received my master's degree. MCC will always hold a special place in my heart. It not only provided a good learning experience, but it also helped give me the financial assistance needed in fulfilling my dream ... to be a teacher.

> *Maria Lopez Miller graduated from McLennan Community College in 1972 with a degree in secondary education and now lives in Waco, TX. She has been a ninth-grade teacher of English and Spanish in the Waco Independent School District's magnet school for 16 years, where she serves as a role model for all students in her school.*

Charles A. Mitchell
Wilkes Community College, NC

Wilkes Community College kindled the latent spark in me that gave me a burning desire to be the best person I could possibly be—no matter what other people say or do. WCC encouraged me to dream and set high goals. Today I am successful because I continue to dream, to set goals, and to reach goals.

For an unmotivated, underachieving high school student, Wilkes Community College played an integral part in developing me into a successful dreamer of dreams, goal setter, and mover of men. I am an experienced educator, human development specialist, community leader, youth sports coach (Tee Ball and YMCA Mini-Mites Basketball), Freemason, father, and husband. After Wilkes Community College I received three higher degrees from Appalachian State University and pursued a doctorate from Virginia Tech.

My public school experience encompasses 1st through 12th grades as a teacher, counselor, and presently an administrator. I am a human development specialist assisting individuals and small companies with goal setting, communications, and decision making. As a community leader I spearheaded the successful two-year effort that saved Elk Creek from sewage contamination by having it designated an Outstanding Resource Water (ORW). I hold offices in several organizations including the Masonic Lodge. Sports have always been important to me. Coaching young people gives me an opportunity to build stronger bodies and stronger minds. Family is always first. I love to have quality time with my family.

Charles A. Mitchell is a graduate of Wilkes Community College now living in Ferguson, NC. He is a human development specialist assisting individuals and small companies with goal setting, communications, and decision making. He is a community leader, youth sports coach, and office-holder in several organizations including the Masonic Lodge.

Alice Ann Moore
Kankakee Community College, IL

After graduation from high school in 1968, I was fortunate to be part of the first class to enroll at Kankakee Community College. I had decided to major in art, but an accident forced me to withdraw from college.

Years later I married, had two beautiful children, and eventually divorced. My friends encouraged me to enroll in Kankakee Community College so I would have the proper skills needed to support my children. The dentist's office where I was employed had no future or benefits. In 1985, I took the first step and talked to a counselor at KCC. I worked full-time at the dental office, attended classes at KCC, managed a home and two children, and had to battle the disease multiple sclerosis.

Kankakee Community College has given me the quality education needed to achieve my goals. My business classes have given me the skills needed to hold the variety of jobs necessary to help finance my education. The interpersonal communication classes have given me the ability to be assertive, yet sensitive in a business world that is competitive and sometimes stone-cold to human relations.

KCC has provided excellent educational opportunities. It has given me more than book knowledge; it has helped me gain self-esteem, confidence, and the ability to take risks. I have learned that maintaining values and principles is as important as profit and dollars.

KCC has given me the basis for a successful education at a senior institution. I have felt very confident to continue with my studies at Governor's State University because of KCC. When I graduated from KCC in 1989, I was awarded the Community College Scholarship to GSU. I am currently enrolled in the media communications program with emphasis on television production. With the skills I achieved at KCC, I am working as a production assistant in the Instructional Communications Center, the college TV studio. I am still a member of the Marketing/Management Advisory Board at KCC, a founding member of the Manteno Cable Access League , a local cable access station. I will be producing video tapes that will be aired on Manteno's Channel 10. I am also employed as advertising director of the *Manteno News*.

I have taken advantage of all opportunities that Kankakee Community College had to offer from their clubs to the educational events. My daughter Kelly is now a freshman at the college.

KCC taught me a very important lesson—you are never too old to reach for your goals. This summer I decided to make a statement about my feelings for KCC. Wearing a KCC sweatshirt, I skydived from 11,000 feet. My statement to all students and potential students is...You Can Reach New Heights at KCC!

Alice Ann Moore graduated from Kankakee Community College in 1989 with a degree in marketing/management and now lives in Manteno, IL. She is a production assistant in the instructional communications center at Governors State University. She is also an advertising director for the Manteno News *and owns a media company that helps small businesses free of charge. As a member of the Multiple Sclerosis Society, she counsels newly diagnosed patients. She also advises newly divorced women and is active in both Girl Scouts and her church.*

Manuela Mosley
Community College of Denver, CO

After I finished high school in Spain, I registered in college with no other motivation than to create a source of irritation for my mother, who was against females getting an education since we were destined only to become wives and mothers.

A year later I met my husband and happily left school and, shortly after, we left Spain for the United States. My feminine duties were being fulfilled: I was a wife and had become the mother of two children. But my marriage ended after five years and I regretted my lack of education. By working two jobs, however, I was able to support my family.

The years went by and my children left home. I was lost, my drive and ambition gone, and I realized how much I hated my occupation as general manager of a restaurant. I started working as a teacher's aide at a high school and became involved in community service. I loved it. It was highly rewarding to share my somewhat limited knowledge with those young people and even more so to motivate them and pique their curiosity. The students as well as the faculty at the high school encouraged me to get a BA and become a teacher. But they didn't know how afraid I was.

It had been almost 30 years since I was a student; I was sure I'd forgotten everything I'd once learned, and I was petrified that my limited English would impede my learning. I found the courage to register at the Community College of Denver, telling myself that if I failed, at least I'd tried.

CCD was the best thing that ever happened to me. I registered in the fall of 1987, a couple of months short of my 42nd birthday. I found many other students my age who, for reasons of their own, were returning to school. We all found a caring faculty and staff. Meeting these older students encouraged me further, but the faculty made my academic success possible.

These dedicated professionals had forgone glory, recognition, and honors as university scholars, and chosen instead to concentrate their efforts on teaching people like me. They built my confidence by bringing to my attention all the skills I'd acquired through life experiences and building upon them. Where there was a gap in academic knowledge, they didn't merely bridge it, but filled it with knowledge so I could walk on firm ground.

I graduated with honors from CCD in the spring of 1989, and I'm presently pursuing a bachelor's degree and teaching certification with plans to continue on to graduate school for a master's degree and possibly a doctorate in language. Thank you, CCD!

Manuela Mosley graduated in 1989 from the Community College of Denver with a degree in Spanish and now lives in Lakewood, CO. She is a full-time student/ teacher in Lakewood and was chosen as CCD's 1989 outstanding graduate. She participated in numerous activities in college, including many that helped non-native English-speaking students.

Thomas C. Nielson
Columbia Basin College, WA

While Columbia Basin College was located near my home, I didn't seriously consider it, since I was caught up in the intrigue of fraternity life at Washington State University.

In the fall of 1960 I eagerly left home and began studies and an active social life at WSU. I enjoyed friends and tolerated large university classes. At the conclusion of my first semester I was somewhat horrified to find that I had only earned a 2.8 GPA, since I had been accustomed throughout high school to being in the top portion of my class. I told myself it was only an adjustment to the new environment and surely during the second semester I would redeem myself. Unfortunately, when I returned for the second semester, the social demands remained high, and the large lecture classes became easier and easier to skip.

Within the first month of my second semester, I became very ill and was forced to drop out of college and return humbly to my hometown. Within weeks the spring quarter at Columbia Basin College was to begin, and since I was feeling better I thought I might as well pick up a few classes before I returned to the university in the fall.

During my first quarter at Columbia Basin, I was very surprised to have three excellent instructors who really cared and who challenged me to perform my best. Having small classes was a real advantage, as it brought out my natural competitive spirit and fostered outside-of-class study groups and discussions with fellow classmates. I surprised myself by actually enjoying learning, and my community college instructors helped me to expand upon my own areas of interest and curiosity. When it came time to make a decision to return to the university or remain at Columbia Basin College, the choice was easy. Clearly I was getting a much better education at the community college than I had experienced at the university.

When I transferred after earning my associate degree, I decided that I wanted to teach at a community college. I approached my studies with a new sense of purpose and was able to complete my bachelor's and master's degree in record time.

I then obtained a teaching position at an in-state community college, and a year later I was selected to an administrative position at brand-new Mt. Hood Community College in Oregon. I am now in my 24th year of community college work and occasionally reflect back on the significance of Columbia Basin College in influencing my chosen profession. I have never regretted selecting community college education/administration as my profession. I am proud of the opportunities our system provides people from all walks of life while remaining true to the high standards of excellence in teaching. Each year at commencement as I congratulate our graduates, I feel a special surge of pride for the expanded opportunities our community college has given to members of our community.

Thomas C. Nielson graduated from Columbia Basin College in 1963 with a degree in psychology and is now living in Edmonds, WA. He is the president of Edmonds Community College and was selected president of the year in 1988 by the Association of Community College Trustees, Pacific Region, among many other honors. He is active in numerous professional and civic orgnizations.

Patrick J. O'Connor
College of DuPage, IL

If you think of your formal education as a long flight that started in kindergarten, the end of high school marks the time when you have to start thinking about how you're going to land the airplane. In my particular case, I didn't know when or where I wanted to land.

I had been accepted to the University of Illinois' architecture program (a five-year, $25,000 endeavor) and was a month away from going (I had even been assigned a roommate—I hope the guy forgave me for not showing up). The problem was that I had just started to develop into a pretty good tennis player that summer, and I wanted to keep playing, but I wasn't good enough for a Big Ten powerhouse like the University of Illinois. I checked out the College of DuPage and discovered that its associate degree would transfer just about anywhere, and its tennis team consistently won the state title and placed well at the national junior college tournament in Florida. I registered, met with the tennis coach, and proceeded to experience two of the most rewarding years of my life.

I had an English composition teacher named Mr. Peranteau, who undoubtedly has no idea that he became the impetus for my desire to write fiction. I also took a religious studies class taught by a practicing Zen Buddhist (properly rotund and jovial), and two business law classes taught by local attorneys. Who could better teach these subjects?

The tennis program was, and still is, phenomenal; it compares favorably to most university programs. Dave Webster, our coach, taught us how to practice, how to improve, and how to win graciously and lose just as graciously. During my sophomore year we won the state title and placed in the top 14 nationally. I had scholarship offers from many four-year colleges, and I accepted the offer from Santa Clara University, an NCAA Division I school.

After graduating with a bachelor's degree in English, I taught tennis at a club for three years while I wrote a monthly column for *Inside Tennis*. The coach at SCU retired that year, and I was hired as his replacement. I coached for one season before I decided to continue my education. I was accepted at the University of Arkansas' master's of fine arts programs in creative writing and was given a teaching assistantship. Right now I am teaching two classes of freshman literature and composition and working on a collection of short stories.

Teaching college freshmen is interesting. I am convinced many of them would be better off in a junior college closer to home. All of them are surprised that I, their beloved teacher of English, attended a junior college. I always tell them that I could have gone to a four-year university right after high school, but I wanted to stay home, fly around a bit, and discover what I wanted to do. I tell them that junior colleges allow you to stay near home while you keep your plane in the sky. Do the work at the junior college; then you can fly and soar. You can do loop-de-loops.

Patrick J. O'Connor graduated from College of DuPage in 1983 with a degree in English and now lives in Downers Grove, IL. He is a college instructor. According to one of his former college instructors, he was a role model, an athlete who also excelled as a student, and a person who represented the College of DuPage with sportsmanship and grace.

269

Charlott Allen Ostermeier
Lincoln Land Community College, IL

When I began my college career, I had been out of high school for almost 10 years. My years in high school were great fun. It was a great socializing experience for someone raised as an only child. I loved learning, but I never realized the importance of studying toward a perfect GPA. In high school I paid attention in class but didn't see the need to study. I graduated fifth in my class of 40. Never did my high school counselor mention attending college.

The state employment service sent me on an interview at Lincoln Land Community College in March 1971. I began working in April and came into close association with co-workers, students, faculty, and administrators who were very involved with higher education in a new school. It rubbed off.

My first two classes were tennis and classical piano—two things I never had the opportunity to learn before. As I progressed through the beginning academic courses, it became addictive. The more I learned, the more I learned I didn't know, the more I wanted to know, and the search for knowledge become even more intense.

By coincidence, the Lincoln Land Community College Alumni Association was founded the year I received my first degree. As a charter member I have remained active by serving as secretary and as first vice president.

I was also a charter member of Phi Theta Kappa (honor society). Although I maintained my interest through the years, my active involvement peaked during the 1988-89 school year, when I served as vice president, as a member of the scholarship committee, and as financial manager during the campaign to elect one of our presidents to a national office.

I became a member of Alpha Beta Gamma (business honor society) in 1986 and served actively in the group from the beginning. I served on the executive board as secretary and as president for two years.

After all these years of school/learning/education/living—whatever you want to call it—I have found that I am a part of a never-ending circle. I have learned because others have shared, and I am happiest when I am able to share what I have learned with others. My volunteer activities include both structured and unstructured events. Some of these events bring so much pleasure that I feel I should pay for the privilege. These events include retail selling for non-profit groups, facilitating workshops, televised fundraisers, and student-teaching assignments. I have been involved with LincolnFest, Capital Sertoma Club, the Ethnic Festival, Parents Without Partners, First Night Springfield, United Cerebral Palsy, the Greater Springfield Chamber of Commerce, Goodwill Industries, the Volunteer Connection, and others.

Charlott Allen Ostermeier graduated from Lincoln Land Community College and received degrees in 1981, 1988, and 1989. Her major subjects were physical education and business. She is a typesetter at Lincoln Land Community College and lives in Springfield, IL. She is involved in many volunteer activities and has contributed to various community organizations.

Norman D. Randolph

Community College of Allegheny County, PA

In the communities surrounding the mill towns in the Mon Valley of Pittsburgh, many people were encouraged to seek jobs in the steel mills rather than attend college. This was particularly true if you grew up as I did, in a project housing complex. Looking back now, I wonder if I didn't value college because several of my older brothers weren't afforded opportunities at college and subsequently ended up serving in the military. Like myself, all are Vietnam era veterans.

With a drive to do more and at the urging of my brothers, I applied to the Community College of Allegheny County, South Campus. In January 1971, I began commuting daily to South Campus as a freshman. The college was housed in hotel rooms, old buildings, and in the basement of a church in McKeesport, not far from my hometown. For the most part, the college was still in its infancy, struggling to survive as an institution of learning. Some people were embarrassed to attend South Campus because the physical surroundings on campus were not like the traditional college campuses.

In spite of these things, I attended South Campus and earned an associate degree without missing any days of school. I knew that the physical structure of the buildings was not as important as the people who taught and counseled in them. The teachers, counselors, and administrators made the difference. That difference nourished my enthusiasm into ambitions and goals that compelled me to seek opportunities in public service. If I were to define the difference, I would call it "The CCAC Philosophy."

At South Campus I learned that teachers and counselors are some of the most wonderful people in the world. The teachers taught me in a way and at a level that I understood. Through goal-setting, positive encouragement, and personal care, I became a motivated force on the road to success. The family at South Campus made me believe that I could achieve and be successful.

This belief became a reality after I graduated in 1972 with an associate degree in business management. Following graduation, I was awarded a Ford Foundation Scholarship for being the top Black student academically in all of the community colleges in Allegheny County. This scholarship enabled me to attend the University of Pittsburgh and succeed in earning a BS, master's degree, and a doctorate in education.

Norman D. Randolph graduated from the Community College of Allegheny County in 1972 with a degree in business management. He now lives in West Mifflin, PA, and is a high school principal. He has written successful proposal programs for At Risk, Sex Equity, and Adolescent Parenting programs with the public schools in the Pittsburgh area.

As a principal of a junior/senior high school, I realize how grateful I am to have had South Campus prepare me for public service. Helping young people to acquire an education and become responsible citizens is a very rewarding endeavor for me. Many thanks to all of you at South Campus. You provided the vehicle I needed to find my way from Vietnam to public education. I will be eternally thankful.

Audrey Ray
Housatonic Community College, CT

I re-entered college after a 22-year absence; I was separated, facing a divorce, and had to make myself marketable so I could support myself and my three children.

After calling the colleges close to my home, I chose Housatonic Community College because it was less expensive than the others and it had a secretarial program. My goal at the time was to gain the skills to get a job as a secretary. On the day I registered for college, I drove up and down the street in front of the college to find the courage to enter; it had been a long time since I had stepped inside a classroom as a student.

But once inside, I found I could do it. I learned to set priorities. I learned how to take notes, to study for exams, and to write term papers. One of my first professors inspired me because she was blind; that she had succeeded made me more determined to succeed. The faculty and the Women's Center provided support and encouragement. At the center, talking to other women students, I learned how to manage time and how to cope with being a mother, a student, and a head of the family. I learned to adjust, change, and adapt.

As I gained confidence in my ability to achieve, I took a work-study job as a peer counselor in the Women's Center. Seeing other women gain confidence as they succeeded was especially rewarding.

One of Housatonic's strengths is its ability to achieve a blend in the student body. There is constant interaction between peoples from diverse backgrounds and generations. I learned about my own children by studying alongside young adults of their age. While at the beginning I had been nervous about being in class with teenagers, I soon found them helpful and friendly. We learned from each other.

In 1984 I earned an associate degree with honors. By then I had changed my goals, and decided to transfer to Western Connecticut State University, where I earned a bachelor's degree in business administration—human resources management. Since then, I've received a master's degree in teaching from Sacred Heart University and started a new career—I am teaching high school students in a special program at Platt Regional Vocational High School.

Looking back at where I was six years ago, I know that I have come a long way. I also realize I have farther to go. There are opportunities of which I have yet to take advantage; experiences that I have yet to know; areas of knowledge that I would like to pursue. Housatonic helped me start my new life and provided me with a great education and a supportive atmosphere.

> *Audrey Ray graduated from Housatonic Community College in 1984 with a degree in business administration and now lives in Huntington, CT. She is a high school teacher who has built on earlier experience as a peer tutor to provide high school and college students with a model of growth, achievement, success, and caring.*

James R. Regan
Massachusetts Bay Community College, MA

Because I attended a small parochial high school in a college preparatory program, it was assumed I would attend college directly after high school. However, during my senior year I decided to work for a year before attending college. My decision was based upon my uncertainties about what college program to pursue and my career goal. In May I changed my mind and applied to various colleges. Since my late decision did not grant me admission to a four-year institution, I am thankful now that I was advised to apply to Massachusetts Bay Community College.

My experience at MBCC was a pleasant, rewarding one. I had applied and was accepted to the business management program. My reasons for applying to business were not clear, for I had never taken a business course in high school. The initial weeks in my first semester were difficult, since accounting and economics introduced new and different concepts that were more than challenging and not comparable to any courses I had completed in high school. However, I decided to try a semester in the business program and mid-semester found I was doing quite well. Having missed the dean's list by one grade, I decided to strive to attain it in ensuing semesters, which I did.

Perhaps defining my career goal and redefining my value system were the two most important things gained from my experience at the community college. First, I could see that there was something gained from every class I completed—if not in the course content itself, perhaps in the instructor's philosophy of life. I realized I could not develop my abilities and attain knowledge just from studying the courses I enjoyed and contending with the remaining courses as I had done in high school. I also decided that some day I would like to support students who were as lost as I was, giving them the same encouragement I received from the faculty, administration, and staff at MBCC. What I wanted to do most with my career was to help others feel and realize the same success that I had at the community college.

Having worked a few years in business, I never expected that the opportunity would arise so early in my career to work in community college administration. I am very thankful to the community college staff, for they gave me the inspiration to set my goals and to achieve my ambitions as well as the motivation to pursue and succeed at anything I wished to attempt.

James R. Regan graduated from Massachusetts Bay Community College in 1969 with a degree in accounting and now lives in Canton, MA. He is an associate dean for admissions and enrollment development and has provided opportunities for the nontraditional student to enter higher education by reaching into the culturally diverse communities of Greater Boston.

Elizabeth B. Rhodes
Virginia Western Community College, VA

When I was a junior in high school, my parents offered to send me to the college of my choice. Not having any idea what I wanted to do, I went to my high school counselor for advice. His advice was to forget college (in his opinion, I was not "college material") and maybe attend a secretarial school. At the time, I was taking both college prep and business classes and was not thrilled with the idea of becoming a secretary. Not knowing what to do, I put my career decision on hold.

During my senior year, I made two major decisions. First, I decided I could succeed in college; and second, due to my new feelings of independence, I would pay my own way through college. This second decision limited my choice of colleges to our local community college, Virginia Western Community College. My two years at VWCC were not only a time of educational enrichment, but also a time of confidence-building. After all, I was advised by a professional to not even attend college! The majority of my instructors were interested not only in teaching their subject matter, but also in each student as an individual. Never before had I experienced such compassion from instructors and so much willingness to help.

After two years, I transferred to Virginia Commonwealth University and experienced what I now know was an extreme case of "transfer shock." I expected an environment like VWCC, and that was definitely not the case. However, I was graduated with honors after two years with a degree in business education. Prior to graduation, I received the National Business Education Association's Award of Merit for the Outstanding Student Teacher in business education. I taught high school in Richmond for one year before returning to VWCC as an instructor. In 1981, I received my master's degree from Virginia Polytechnic Institute & State University in community college education.

This is my 14th year of teaching at VWCC. As each year passes, I see how much more we are fulfilling the mission of the community college. Our goal is to help our students become achievers—from the high school graduate who is a first-generation college student, to the mother who has raised her children and is now seeking education to better her employment opportunities, to the father who never had a chance to go to college due to family obligations, and to the retirees who simply want to learn. Although the difference in characteristics of the people provides a great diversity in the classroom and thus a challenge to the instructor, the students all have one thing in common: the need for encouragement and understanding—the same kind of help that I received many years ago. I believe we give just that at VWCC, and I am proud to be one of the educators in the community college system.

Elizabeth B. Rhodes graduated from Virginia Western Community College in 1972 with a degree in data processing and now lives in Roanoke, VA. She is an assistant professor at Virginia Western Community College, thereby serving as a positive role model for many men and women. She has performed volunteer work with children's programs at the YMCA and is also a member of VCCA.

Raymond L. Rickman
Okaloosa-Walton Community College, FL

I graduated from high school in 1975. Feeling pressure to obtain employment, I chose to forgo college and enlisted in the Air Force.

Upon being assigned to a duty station, my intentions were to further my education while in the Air Force. To my disillusionment, every avenue along which I tried to pursue my education closed at some point.

Near the end of my four-year enlistment, I decided that I would not re-enlist, but rather I would leave the Air Force and attend Okaloosa-Walton Community College, FL. I chose OWCC because I had been able to take some college-prep courses there and because academically it was one of the highest-rated colleges in Florida.

I entered OWCC in June 1979 and completed my associate of science degree in drafting in 1981. The semester prior to graduation, David L. Goetsch, my drafting instructor, lived up to his promise of placing me in a job. He also persuaded me to further my education by promising to help me find a teaching job when I received my BS. Being the man that he is, he again lived up to his promise. As a result, I am presently lead instructor for drafting at OWCC.

As I reflect back on my days at OWCC, I realize that my community college experience has been extremely beneficial. It has enabled me to acquire the skills needed in preparation for employment. It has taught me life-long learning skills that I use on a daily basis. But the thing I remember most about college is the instructors, especially my drafting instructor. He showed a sincere desire in helping me achieve my goals. His help extended beyond the normal academic instruction; he paid personal attention to my needs, and we developed a friendship that is still perpetuated today.

The vocational education I received at OWCC was an outstanding bargain for the price. The extra attention I received has been instrumental in molding my attitude and feelings toward teaching. I, like my mentor, seek to develop an instructor/student relationship of mutual respect, one of sincere concern for students' well-being. I feel very fortunate to have received my initial education from OWCC and hope that the thousands of students who enter the community college environment find it just as rewarding.

Raymond L. Rickman graduated from Okaloosa-Walton Community College in 1981 with a degree in drafting and design and now lives in Shalimar, FL. He is a drafting instructor at Okaloosa-Walton Community College and has written books in his field. He has set an example of dedication and perseverance for all to follow. He was honored as an outstanding alumnus at the college's 25th anniversary this year and has presented numerous professional workshops on drafting.

Robert J. Romine
Independence Community College, KS

When I entered the Independence Community College freshman class in the fall of 1965, I became the first member of my family to attend college. Unlike many of my high school classmates, I had to hold a full-time job. Balancing the roles of full-time student and full-time employee was not easy, but with the help and encouragement of my family and my instructors, I managed. The accessibility of the community college, the flexibility of the institution in meeting student needs, and its sheer user-friendliness provided me an opportunity to launch a successful academic career.

High academic standards were enforced during my community college experience. Rigorous testing and recitation were the rule. Communications skills were taught in all courses, and proper grammar, spelling, and punctuation were expected in all classes. Yet, caring instructors were always available for assistance, and personal and academic counseling were commonplace. With these structural advantages, my community college instructors began to work that special brand of magic that only gifted teachers can perform. The wonder of learning new things, of gaining fresh insights, and of mastering new skills began to germinate. Soon it grew into a fully developed excitement and enthusiasm for learning that would serve me well throughout my academic career.

I would be remiss in failing to note one other characteristic of my community college experience that was and still is invaluable to me. My exposure to higher education at Independence Community College emphasized similarities among disciplines and skills rather than dissimilarities. This resulted in an impression of the interconnectedness of human experience and the sense that all disciplines dealt with different aspects of the same greater subject. It left with me the distinct sense that the connections between and among bits of information, facts, or events were often far more significant than the bits themselves. My community college experience emphasized integration and synthesis of knowledge.

During my undergraduate university years, community college education became my career choice. My decision was a conscious one, derived directly from my own community college experience. Having been an honor graduate and a teaching fellow, and after completing two graduate degrees, I secured a community college teaching position and have been involved in community college education ever since. I was honored by my colleagues as a master teacher and served as a division chair, assistant dean of instruction, and a community college specialist with the Kansas State Department of Education. Today, I ensure students the unique benefits and special learning environment characteristic of the community college, while serving as the dean of instruction of my alma mater.

Robert L. Romine graduated from Independence Community College in 1967 with an associate's degree and now lives in Independence, KS. He is the dean of instruction at Independence CC and has been active in the Kansas educational system for many years. He was chosen as Outstanding Educator of America in 1973, 1974, and 1975. He assisted in the design, editing, and review of various programs adopted by the Kansas State Board of Education and has participated in activities of the Kansas Committee for the Humanities.

Jo Ruta
Chattanooga State Technical Community College, TN

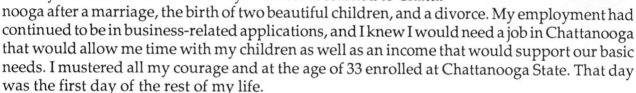

My education began in a one-room schoolhouse in a rural farming community in Milo, TN. My one-room country school had only one teacher, did not possess state accreditation, and did not award grades. At the age of 10, when my family moved to the city, I was assigned to the second grade. During my next school year, I completed the third and fourth grades, and the following year grades five and six. After high school graduation, I enlisted in the U.S. Air Force and was assigned duty in Europe.

After serving my tour of duty, the urge to see more of the country sent me to New York. Six years later I returned to Chattanooga after a marriage, the birth of two beautiful children, and a divorce. My employment had continued to be in business-related applications, and I knew I would need a job in Chattanooga that would allow me time with my children as well as an income that would support our basic needs. I mustered all my courage and at the age of 33 enrolled at Chattanooga State. That day was the first day of the rest of my life.

I majored in accounting and data processing, continued my education at the local four-year university, and received my undergraduate degree in business education in 1972. I began my teaching career at a local high school, continued to pursue my education, and received a master's degree in administration and supervision in 1975; within another year I completed the requirement for a master's degree in accounting and data processing in education. While teaching at the high school, I was a part-time faculty member at my two alma maters, Chattanooga State and the University of Tennessee at Chattanooga.

A high point in my career occurred in 1979 when I returned to Chattanooga State as a full-time instructor in the computer science department. Some of my former professors, who had given me so much needed encouragement, were still teaching at Chattanooga State. During my college years, I was faced with many difficult decisions, financial worries, and my daughter's major medical problem. The faculty at Chattanooga State gave me the encouragement and support that I needed to continue my education. I try every day to return to my students the thoughtfulness and consideration given to me when I so desperately needed it.

During the past 10 years, I have continued with my education. I will always be a student—I love learning and teaching. My profession has given me many opportunities to grow.

Jo Ruta graduated from Chattanooga State Technical Community College in 1970 with a degree in accounting and data processing and now lives in Chattanooga, TN. She is an associate professor of information systems and has been appointed by the governor to serve as the Tennessee chairperson for the Developmental Disabilities Planning Council. She was also selected as a volunteer consultant in a national project designed to assist elementary and high schools with the installation of computers.

Elaine Ryan
Rockland Community College, NY

As an entering freshman at Rockland Community College in 1974, I did not realize that my aspirations would be fulfilled for many years to come. Nor could I have perceived that 13 years later I would return to Rockland Community College as an assistant professor in the business department.

During my first undergraduate year, my English professor suggested that I participate in the college's London Honors Program Abroad. I anxiously accepted the honor to study for a semester in London since I had always been intrigued by the prospect of living in a cultural environment similiar yet very different from the one I knew. Visits to Stratford-on-Avon, Canterbury, and Stonehenge brought to life the literature I had studied. The depth of my understanding was greatly enhanced by being able to compare and contrast American and English norms. My positive experience in England increased my desire to discover other cultures, and I eventually traveled to France, Italy, Greece, the Greek Islands, and Israel.

As an RCC professor, I try to expand my professional and teaching role to encompass global issues, which will enable me to offer the same encouragement and enrichment to RCC students that I received. My accomplishments as an RCC professor have been most rewarding. In April 1989 I was selected as an awardee under the U.S. Department of Education's Fulbright-Hays Seminars Abroad: China's Economic Development. It is the first program devoted specifically to economics and economic reform in the People's Republic of China. During the spring of 1989, I was actively involved in the coordination of a professional development seminar in Belgium regarding EEC 1992. The seminar participants included small business entrepreneurs and faculty members from State University of New York and City University of New York community colleges.

In anticipation of continuing my international learning experiences, I am presently a committee member for the administration of a Matsushita Grant for Development of Faculty Awareness of Japanese Culture at Rockland Community College.

I am looking forward to pursuing a doctorate in international business. Thank you, RCC, for fulfilling my aspirations for many years to come!

> *Elaine Ryan graduated from Rockland Community College in 1977 with a degree in liberal arts/English and now lives in Cornwall, NY. She is an assistant professor at Rockland Community College and serves as an educational planner there for a variety of students. She won a Fulbright Seminars Abroad Fellowship in 1989.*

L. Diane Ryan
Cuesta College, CA

Cuesta College has figured prominently in my life, from the beginning of my association there as a floundering student in 1968 to the present day, as we in higher education are preparing to move into the '90s with an eye to the ever-important role played by the community college.

In some ways, my educational resume reads like the blueprint for the California Master Plan for Higher Education, inasmuch as I obtained the associate degree from a community college, transferred to a state university, where I completed a bachelor's degree, and later went on to the University of California, where I am currently a doctoral candidate. However, Cuesta College was my second stop in the postsecondary experience and an all-important one for me.

I began college life as a Regents Scholar at the University of California but soon rejected the "establishment" offer and very nearly rejected college altogether. The turbulent events of the '60s took their toll on all of us in different ways. Cuesta College provided me with a second chance rather than a forced diet of scholarship—and I was permitted to come to the educational table as a hungry participant.

Through interaction with dedicated faculty and administrators, I learned about the academic community. In working with community members to create a permanent campus for Cuesta, and in the tremendous variety of other community services sponsored by the college, I learned about community in the larger sense.

Community colleges have greatly expanded the opportunities for access for thousands of students. Making the transition to a four-year institution is often difficult and made more so due to financial difficulties. My career has been chiefly devoted to the many aspects of financing postsecondary education for students and families. As a financial aid administrator in a public university, and through my work with several non-profit corporations that promote higher education, I see the all-important role community colleges play in expanding opportunity.

Despite the enormous challenges facing higher education, I know that the community colleges are where a real difference can be made in the lives of those served. I am proud to have obtained an associate degree from Cuesta College and know that my experience there had a profound impact on my personal and professional life. Cuesta College gave me a chance at life, and I am grateful to be a part of the higher education community that makes similar chances available to thousands of students.

L. Diane Ryan graduated from Cuesta College in 1969 with a degree in social science and now lives in San Luis Obispo, CA. She is associate director of financial aid at California Polytechnic State University and a member of the Board of Regents of Cathedral High School in Los Angeles. She has also served as consultant to Sigma Systems, Inc., a financial aid software company in Los Angeles. Her professional volunteerism includes extensive work with the California Association of Student Financial Aid Administrators, and she has also served as a member of the Board of Directors of the California Higher Education Loan Authority.

Mioko Saito
Rose State College, OK

It all started as revenge on the Japanese higher education that rejected me. I was rejected for a legitimate reason; I did poorly in calculus, geometry, algebra, trigonometry, physics, biology, chemistry, geology, world history, Japanese history, political science, geography, modern Japanese, classical Japanese, classical Chinese, and modern English, to name a few. As a consequence, I gave up college; instead, I went to a business school for two years and worked at Mitsubishi Heavy Industries for three years, which together prepared me for going to an American college.

However, because of my poor high school grades, I was rejected by the University of Oklahoma, although my TOEFL score was high enough.

I was very frustrated with my ability and with the educational system as a whole when I started going to Rose State College in 1981. But surprisingly, I became a hard-working student. Regardless of my background, my teachers always believed in me and encouraged me. I not only finished the minimum requirements for acquiring my associate degree in business, but also I found more potential in myself. I learned how to learn and improve myself. I learned the joy of making good grades.

It was only a year and a half after my arrival when I received the acceptance letter from the University of Oklahoma. I actually breezed through my undergraduate years. When I went through the graduation ceremony with honors in 1985, I was already a graduate student pursuing a master's degree in education.

Now I am at the dissertation stage of my doctoral degree in educational technology, planning to finish in May 1990. My motivation and goals have changed from revenge to a new hope for education in general. If I can make it, anybody can. Rose State College taught me a very valuable lesson: It is not too late to start if you are willing to learn. Rose State College gave me the second chance that I would not have had otherwise. When I go back to Japan with my Ph.D., I would like to let Japanese high school students fighting college examinations know that it is not the end of the world. I would like to encourage them just as my teachers at Rose State College encouraged me.

> *Mioko Saito graduated from Rose State College in 1982 with a degree in business and now lives in Norman, OK. She is at the dissertation stage of a doctoral program. She has inspired her fellow classmates with the high standards she has set and attained as she pursued a college education in a language other than her native tongue. Evidence of her abilities is exemplified by the fact that the United Nations has offered her a position as an educational consultant for Third World countries.*

Cecilia Elaine Salisbury
Thomas Nelson Community College, VA

Being a winner is an attitude that you carry deep inside your heart, but finding a place that encourages you to strive for your best is an opportunity that all too many people fail to recognize. The Virginia Community College System is set up for those who need a place to get started, and Thomas Nelson Community College should pride itself on its accomplishments in this area. I started my first full year at TNCC not only with some rather severe physical limitations, but also after having been out of school for 20 years. Yet, I graduated in May 1989 with highest honors and was chosen that year as one of 10 national finalists for the title of Distinguished Student Scholar by the AACJC and Phi Theta Kappa.

Now I am an undergraduate in the Honors Program at Virginia Commonwealth University. I received one of the first full Presidential Scholarships to be offered by VCU. My experiences at TNCC prepared me for successful entrance into this university. My classes at TNCC gave me the kind of background necessary to feel at home in all of my present classes, and being editor of the student newspaper was a priceless, irreplaceable experience.

There is nothing second-rate about the education offered at community colleges. The opportunities I had for service to my fellow students with the newspaper and as a tutor gave me confidence, but I was also given the chance to build strong and meaningful relationships not only with other students, but also with the faculty, staff, and administrators. My own actions made things happen, but it was the encouragement and assistance of the faculty, staff, and administrators of Thomas Nelson that kept me going. Their belief in me and my abilities and their willingness to provide me with the opportunities to prove myself allowed me to achieve a level of excellence I had only dreamed about before coming to Thomas Nelson.

You might say that this is all well and good for someone who is self-motivated, but how about someone who doesn't really know what he or she wants to do or where to go? I would highly recommend the community college for this situation. There is an opportunity to explore and a dedication to the individual student that you will not usually find in a larger institution. Age and gender are not barriers. You don't have to be just a number, and you are as important as you make yourself. You can get involved in activities or you can watch from the sidelines. You have the freedom to be yourself, and the options are limited only by your own dreams. At a community college, it is all up to you. You make it happen, and you have the full support of a college that wants you to succeed.

Cecilia E. Salisbury graduated from Thomas Nelson Community College in 1989 with a degree in humanities and social science and now lives in Sandston, VA. She is currently an undergraduate student in the Honors Program at Virginia Commonwealth University. She was editor-in-chief of Thomas Nelson's student newspaper, chair of various PTK committees, and an outreach speaker promoting education and stay-in-school programs in local public schools.

Napoleon N. Sanchez
Berkshire Community College, MA

Berkshire Community College played a significant role in my academic pursuits—and in my entire life. I had come to the United States from a village in the Ecuadorian Andes to attend a private school in Massachusetts where I had received a scholarship. When faced with the choice of returning home or continuing at an American college, I decided on BCC. It was a difficult time—I was still learning English, I had to support myself, and I was homesick. And then my apartment and all of my belongings were lost in a fire. But one of my BCC teachers, Arthur Phinney, took me in to live in his family's home so that I could continue my studies at BCC. I graduated with a degree in liberal arts and then went on to receive my BA from the University of Massachusetts.

While pursuing my graduate studies at U-Mass, I began to teach Spanish at Westfield State. In 1977 I earned a doctorate in Latin American literature and was elected to the National Honor Society of Phi Kappa Phi.

I am now a full professor in the Department of Foreign Languages and Literatures at Westfield. I'm also active in such professional organizations as the Modern Language Association, the American Association of Teachers of Spanish and Portuguese, the New Council of Latin American Studies, and the American Council on the Teaching of Foreign Languages.

I'm also a community and global volunteer: I assist with the reading program in the Springfield Public School system; I monitor volunteers teaching Spanish to local children; I translate letters for Amnesty International; and I'm an activist with the Springfield Area Central American Project.

I owe my success to BCC and its faculty for introducing me to the world of higher learning. I owe special thanks to Professor Art Phinney and his family for lending me a helping hand at a crucial time in my life. And I owe my family happiness to BCC as well—I married my BCC sweetheart!

Napoleon N. Sanchez graduated from Berkshire Community College in 1966 with a degree in liberal arts and is now living in Westfield, MA. He is a professor at Westfield State College and has made a difference in the lives of his many students. He received a Distinguished Service Award from Westfield State College and a National Endowment for the Humanities fellowship at Yale University. His commitment to human rights throughout the world has involved him in Amnesty International and the Springfield Area Central American Project.

Sandra L. Scott

University of Maine-Orono-University College, ME

I t was 1973. I was 29 years old and a single parent of two small children. I was on Aid for Families with Dependent Children and had spent the previous two years beginning my recovery from alcoholism. One day in the early spring I saw a brochure from Bangor Community College (presently, University College), which described an associate's degree Mental Health Technology Program. In spite of my lack of confidence and self-doubts, I applied to the college.

I was accepted as a full-time degree candidate, and in September I began classes. Although I was in a Mental Health Program, Mary Lou Cormier, chair of the Human Services Department, knew of my interest in alcoholism and coordinated field placements for me in local alcoholism treatment agencies.

Near the end of the two-year program, Dartmouth Medical School was accepting applications for a training program to prepare alcoholism counselors. Dr. Cormier contacted Dartmouth and asked if they would consider me. I was interviewed and accepted into the Dartmouth program. In June 1975 I received an associate's degree in mental health technology, and in December of 1975 I received a certificate in alcoholism counseling from Dartmouth Medical School.

While at Dartmouth, I applied for and received an appointment as a teaching associate at Bangor Community College. As a teaching associate, I assisted in the coordination and supervising of field placement students in the newly developed program in chemical addiction. In 1976 I was appointed as a special instructor and began teaching the courses in the Chemical Addiction Program. In 1977 I was appointed instructor with the commitment that I would pursue a master's degree. In 1981 I received a BA in psychology and was promoted to an assistant professor, and in 1983 I received a MS in human development and was promoted to an associate professor with tenure.

The academic and career successes that I have just described happened because of my beginning at the community college. It was there that I was provided with an excellent educational foundation, and it was there that I was encouraged and supported and told that I had the potential to be successful. It was at that college that I began to believe in myself.

As part of my role as a faculty member I am active in community service as the coordinator of the Mothers Against Drunk Driving Program. Now I have the opportunity to serve as a role model for other women who enter the college. Some of these women are recovering from alcoholism; many are single parents of small children and have little means of support. Now I am in a position to give to them what was given to me. I am in a position to help them believe in themselves.

Sandra L. Scott graduated from University of Maine-Orono-University College in 1975 with a degree in mental health technology and now lives in Bangor, ME. She is an associate professor of human services at University College and is the coordinator of a Mothers Against Drunk Drivers program. She also serves on boards of directors of substance abuse programs and programs for senior citizens.

Gary O. Seabaugh
Johnson County Community College, KS

Success did not come easily for me. After an unsuccessful semester at a local junior college, I joined the U.S. Navy in 1966 and eventually served in combat with the U.S. Marine Corps in 1968. When I returned from Vietnam, I enrolled at Johnson County Community College.

I have a special fondness for JCCC; it was a wonderful place to be. The faculty members were compassionate and very interested in our learning. They taught relevant and interesting courses and were a major influence in my academic life. I experienced an academic revelation at Johnson County Community College. When I achieved my first "A" on a test I thought, "This is easy; this is fun!" I did not take this revelation lightly because one of my high school counselors had told my mother that I was learning-disabled.

I was a 23-year-old combat veteran. Times were not good for Vietnam veterans then, but JCCC students, faculty, and administrators were enormously supportive. One day I wore my Marine Corps uniform jacket to school—ready for a fight—but the faculty members were supportive of my military service and influenced their students to also be understanding. That compassion was important to the mental and emotional well-being of the returning soldiers and other students.

I was a member of JCCC's first graduating class in 1971. Following graduation I received a BA in philosophy and a BA in counseling psychology from the University of Missouri-Kansas City. I earned a doctorate in developmental and child psychology from the University of Kansas.

In 1978 I developed an alternative high school for drop-outs and students who were expelled from traditional high schools. Today, the Plaza Academy is an accredited high school that specializes in college preparation and special education services through an individualized academic and counseling curriculum. The Plaza Academy provides a junior college environment at the high school level to give students maximum freedom, which encourages responsibility and the acquisition of academic, social, and emotional skills. Seventy percent of our graduates continue their education in community colleges and universities. All of our students are prepared for success once they leave the Plaza Academy.

Just as important as academic accomplishments, the Academy has produced hundreds of students who have gained self-confidence and a positive self-image. We create an environment that gives kids the opportunity and skills to succeed; we teach them to manage their own behaviors. I'm a surrogate father to my students. I want them to feel love and respect from their teachers and parents alike.

> *Gary O. Seabaugh graduated from Johnson County Community College in 1971 with a degree in sociology and now lives in Leawood, KA. As director of the Plaza Academy, he has contributed to the education of hundreds of young people. He was awarded a Purple Heart medal with two Gold Stars and the Silver Star medal for his combat service in the Vietnam War. He has presented his research and model for education at national conventions and was presented the Award of Merit by the Association for the Advancement of Behavior Therapy.*

Fab Snage
Schoolcraft College, MI

I am what they call a nontraditional student, which is a polite way of describing an older student. I attended school in the Middle East while I was growing up. I was a very average student, mostly "B's" with a few "C's" sprinkled in here and there. I had potential, but I was never pushed to achieve. I was a girl, after all!

My husband is American-born. He has been the driving force behind my returning to school. When our daughter turned eight, we decided it would be a good time for mom to start school. I attended a seminar at Schoolcraft College for returning students. The speakers were mostly women, three of whom had already graduated from Schoolcraft. I was very impressed and made an appointment to see the counselor.

In the fall of 1984, 13 years after I left school, I became an official student of Schoolcraft College. I have to admit I was scared to death! To my great relief I discovered that Schoolcraft had a great support system, the Learning Assistance Center. It offered tutorial services as well as classes in note- and test-taking, and I took full advantage of this great service. The instructors were another part of the support system; they were willing to help when needed. My counselor also paved the way for me by recommending fairly simple classes to start out with. She knew I had potential, but she wanted me to like my new experience first; the hard work would come later. I did not disappoint her or myself; I graduated from Schoolcraft with a 3.9 grade point average and an associate's degree.

Schoolcraft had a great impact on my self-worth as well as on my self-confidence. It made me feel that I can do anything. I took four years to achieve a two-year degree because I wanted to go on with my roles as a wife and mother. I also did not want to give up my volunteer activities, and I had to back-track and take some remedial classes. While going to Schoolcraft, I applied for and received a merit scholarship, which not only helped financially, but also was a great morale-booster. I was accepted in the fall of 1989 to the School of Business at the University of Michigan-Dearborn. I also received the Community College Transfer Scholarship from the University of Michigan.

Two to three years down the road, I will receive my bachelor's degree in business from the University of Michigan. I am very confident that I will go on to receive my master's degree. Schoolcraft has turned me on to education, and I am very grateful.

Fab Snage graduated from Schoolcraft College in 1989 with a degree in science and now lives in Plymouth, MI. She is pursuing a bachelor's degree and has excelled academically while still active in the Girl Scout Council, her church, the Plymouth Symphony Orchestra, and Phi Theta Kappa.

Debra Sprague
North Idaho College, ID

When I was in the ninth grade, I had an English teacher who seated students according to their grades. The smart people sat in the back and the not-so-bright students sat up front. . . .I had a particularly bad week soon after the class started and I was moved up to the second row. The teacher started using me and my papers as an example of how not to write a composition. . . .I just gave up completely."

I cringe when I read these painful words in the essay of one of my composition students at North Idaho College. But the writer goes on to say that she gradually overcame her fear and dread of writing because of the encouragement she received from her instructors at our community college.

Another of my students relates in an essay that she dropped out of high school in the tenth grade. Now, 10 years later, she is back, working on a nursing degree. These two students will succeed. Others may not. But these students and others like them would not have the chance to go to college and make up for earlier educational deficiencies and failures, if it were not for community colleges like NIC. It is for this reason that I chose to return to teach at the community college where I spent my first two years of college.

I continued that education at Eastern Washington University and the University of Washington and taught undergraduates at both institutions as a graduate teaching assistant—but I strongly believe that I received no better instruction at these universities than at NIC and that I have taught no more challenging, gratifying, and talented students than those I encountered at this two-year college.

Ten years ago my English instructors at NIC encouraged me to write and introduced me to new literary vistas: T.S. Eliot, William Faulkner, Flannery O'Conner, and the multitude of contemporary women writers who are the mainstay of my reading today. And today, NIC students make me feel that I make a difference in their lives. Most middle-class, traditional students at a four-year university will succeed; one instructor won't make that much difference to them. Here at NIC, the older returning students and those students who weren't "college material" for the four-year schools struggle and are grateful for the concern of a caring instructor. Not all of them will make it, but every day that I see them continue that struggle against the odds—families to care for, full-time jobs, sick children, sick parents, lack of financial support—I am encouraged and proud that I can help them succeed. One more skill acquired, one more paper written, one more class completed, moves them closer to their lifetime goals.

Debra Sprague graduated from North Idaho College in 1980 with a degree in English and now lives in Coeur d'Alene, ID. She is an English instructor at North Idaho College, thereby affecting hundreds of students who might not otherwise be reached by such a sensitive, caring, and adept teacher. She empathizes with students who must surmount many obstacles in order to realize their dreams.

Morris O. Thomas
Northwestern Michigan College, MI

In 1961 I was a senior at a small rural high school (fewer than 200 students) in Michigan. Several of my teachers encouraged me to apply to college, but due to a lack of initiative and some fear on my part, I never got around to sending off the application. When fall term started, I was working on temporary contract with the Forest Service, planting pine tree seedlings for minimum wage. The rest of my time was spent finding part-time jobs, playing basketball, working on the farm, and avoiding higher education.

After one very frustrating day at work, I decided that college might be for me after all. The next day I drove my precious car the two-hour drive to the campus of Northwestern Michigan College. I chose NMC because the basketball coach had sent me some literature and said "hello" the previous year. No scholarship was mentioned, however. I don't remember the subsequent orientation, but everyone was friendly and helpful. I was impressed because there were three or four Black students in the student body.

The English, biology, math, and agriculture classes were very different from high school; the level of expectation was much higher. I survived the first term and earned a small scholarship (I tell my sons now it was a full-ride). I made a lot of new friends and started to see a larger world around me. Several of the faculty recognized my economic plight and helped me get part-time jobs, such as modeling for the life drawing class, coaching elementary school basketball, and washing dishes in the cafeteria (free food!).

During my second year at NMC, I became increasingly involved in classes because they seemed interesting, not just because they were scheduled between 10 a.m. and 3 p.m. The thought of actually continuing for a bachelor's degree became real and I transferred to Michigan State University. After completing an undergraduate degree in soil science, I returned to MSU for a master's degree in geography.

Since then, I've been teaching geography at Lansing Community College, where I see many students who are starting out like I did: searching! Northwestern Michigan College was the right place at the right time for me.

Morris O. Thomas graduated from Northwestern Michigan College in 1964 with a degree in geography and now lives in Lansing, MI. He is an instructor at Lansing Community College. He not only teaches world geography, but also has co-authored a handbook, The World at Your Fingertips, *and has prepared maps for two other books. He has internationalized his curriculum by taking student groups on trips to Belize, Honduras, Guatemala, Brazil, and Argentina.*

Diane Thorpe
Richland College, TX

I'm sitting in my office and reflecting that this has been a usual day for me—which means it's been unusual. One of the things I love most about my job as a community college counselor is that each day brings new facts, new challenges, and new opportunities. Today I taught my class on interpersonal relationships and listened to my students tell me how they had benefited from doing their required special project. I was gratified, as I am every semester, that each student had gained an understanding of what I consider to be the central focus of my class: the ability to alter their behavior for the better. I have taught them a skill that they can use the rest of their lives.

Students are often frightened and confused, but also eager and excited at the same time, just as I was 15 years ago when I started at Richland College. I'll never forget the day I finally got up my courage to call Richland's Admissions Office. I remember asking, "Is it too late to register for this semester?" The person answering assured me it wasn't, and encouraged me to come in right away. I've often reflected that I probably wouldn't be sitting here right now in my role as counselor if the person answering the phone that day had been rude or even indifferent.

I went to late registration, and many classes were filled. Someone suggested a human development course that was still available, and I also discovered a P.E. class that was open—bowling.

I had many fears that first semester. I imagined I would walk into my classes, and everyone would burst out laughing and say, "What are you doing going to college? You're too old!" Or, I'd take the first test and fail, and the teacher would say, "I'm sorry, Diane, but you just don't have what it takes." Or I would try to study and find out that while I had been so busy being the perfect wife and mother, my brain had turned to grey fluff. Thankfully, none of those catastrophic fantasies occurred.

I spent a wonderful three years at Richland, getting to know everyone from the president to the maintenance staff. Each person seemed imbued with the same spirit of warm helpfulness. I learned French, grew fruit flies in biology class, developed my athletic ability, improved my voice and diction, delved into anthropology, sociology, and psychology, and in general, behaved like a kid in a candy shop.

Finally, I had to leave the nest, having received my associate degree. I transferred to North Texas State University, where I collected two more degrees, and later reached my goal of being able to sign my name: Diane Thorpe, Counselor.

Diane Thorpe graduated from Richland College in 1975 and now lives in Dallas, TX. She is a community college counselor and professor at North Lake College with expertise in interpersonal relationships, communication skills, stress management, assertiveness, the Myers-Briggs Instrument, and cognitive therapy. She is a member of the Association of Psychological Type and the American Association for Counseling and Development.

Ron Towery
Spartanburg Technical College, SC

After graduating from high school, I started my freshman year at a local liberal arts college. Having received a scholarship, this seemed to be the right choice for me. I had not committed to a major course of study because I was unsure what I wanted to do for the rest of my life. During my first semester, the company that employed my parents had to reduce their work force and my family was affected by the reduction. I decided to quit college and take a job working in a textile plant to assist my parents financially. During the winter months, I worked the evening shift and also worked a part-time job in the morning. It was not until the spring that I realized that attending Spartanburg Technical College was what I really wanted to do. I would be able to keep my evening job and earn an associate degree in electronics during the day.

While attending STC, I was surrounded by faculty and staff members who were willing to assist me not only academically, but with career placement as well. By attending a small technical college, I received the attention I needed to improve my self-confidence and I had the opportunity to be instructed by role models of the career field that I had chosen. The faculty members provided a positive atmosphere and encouraged me to explore new ideas. At the same time, they told of experiences while working in the field that prepared me for the challenges ahead. My career choice after graduation was in biomedical technology. STC had prepared me for this position by providing me with a solid background in electronics.

One of my goals while at STC was to give back to the college a part of what they had given me—support. After working for five years in electronics, I returned to STC as a part-time faculty member in the electronics area. The position later became full-time. The college still had the same philosphy—to provide a positive atmosphere for learning and to provide support to its students. The support of fellow faculty members provided me with the strength to continue my education and later receive my BS.

Now, as department head of the electronics program, I will continue the tradition of providing support to students in a way that only a technical college can. I realize now how grateful I am for the education that I received at STC. It is much more than a technical college, it is a part of my life.

Ron Towery graduated from Spartanburg Technical College in 1977 with a degree in industrial electronics and now lives in Duncan, SC, where he is an instructor and department head at Spartanburg Technical College. He is active in church, the fire department, and civic clubs. He has served on a statewide technical education task force in integrated manufacturing. His work was cited as an exemplary program by Title III, and he served as interim Title III coordinator for quality assurance at Spartanburg Tech.

Cherylynn F. Tsikewa
Community College of Denver, CO

When I enrolled at the Community College of Denver during the spring 1982 semester, I had only one thing in mind—to brush up on the secretarial skills I already had so I could find a job to support my two children. As a single parent, I had to do something right away to avoid going on welfare, which was my absolute last resort.

During my first semester, I was able to challenge the beginning secretarial courses with the skills I already had, which enabled me to get through the secretarial program at a faster pace. After I completed the first semester, I felt confident enough to continue in the fall. The faculty in the secretarial program were all very understanding and encouraged me to finish the program and get my associate degree.

As I was doing some coursework in the secretarial lab one day, one of the deans came in and asked the instructor if she could recommend a student to work in his office as a typist. She looked around the room and pointed to me. That was how I landed my first job at CCD. After I started working, I found that the staff were just as caring as the faculty. My new boss encouraged me to finish the degree program and even gave me some time off during my work hours to take courses. Without the support and encouragement of my boss and my co-workers, I would have never been able to finish the degree program.

To earn extra money, I tutored residents of an apartment complex for single parents in a new word processing program developed by CCD. I was able to share my experiences with some of the young mothers and encouraged them not to give up on their education. I showed them that it was not impossible to get an education, work a full-time job, and be a full-time mother.

It took only a couple of years to finish school and find a job, but without the support and encouragement from the caring CCD faculty and staff, I would never have been able to achieve as much as I did. I am always very proud to tell people how I came to CCD. I have been here at CCD for more than eight years as a student and an employee and I am still learning new things.

> *Cherylynn F. Tsikewa graduated from the Community College of Denver in 1983 with a legal secretarial degree and now lives in Denver, CO. She is a senior secretary and valuable member of the college staff at the Community College of Denver. Cherylynn is also active in Indian affairs in the Denver area.*

Jerry L. Turnquist
Elgin Community College, IL

For many college students of the 1960s, education meant demonstrations, protests, and general campus unrest. This environment, coupled with burgeoning enrollments at many of the nation's colleges and universities, created a situation in which even the most serious of students had difficulty in achieving a quality education. There were institutions where a different type of atmosphere prevailed, however, and at which students received an education that truly impacted their lives. One of these schools was Elgin Community College, and I am proud to say that the two years I spent there truly changed my life.

To understand the effect that ECC had on me, it is necessary to consider the type of person I was as an incoming freshman. Marginal high school grades and a low ACT score had combined to erode my self-confidence and give me the feeling that my college career would be a short one. When I was required to enroll in a remedial English class while most of my high school classmates took the traditional English 101, I was convinced that I did not have the skills necessary for success. Little did I realize that this was a very fortunate occurrence.

It was in this class that I met Irma Davis, who was undoubtedly the finest instructor I have ever known. She taught me techniques of composition that made me view writing as a pleasure and not a chore. Together with the interest sparked in me by my American history teacher, Karl Lehr, I developed such an interest in local history that I would later be involved in several publications about Elgin history. My involvement on the boards of various local organizations and several community service awards I have received are all a direct result of experience I had during my community college years.

Some might say that I was at a point in my life when any college would have made an impression on me and that it was my maturation, not the teachers, that made the difference. This is most certainly not the case, however. My community college teachers were dedicated individuals who took the time to care about me, and their influence made a dramatic effect on my life. I will always remember them.

Jerry L. Turnquist graduated from Elgin Community College in 1969 with a degree in science and now lives in Elgin, IL. He is a public school teacher and has been involved in producing several publications related to Elgin's history. He is currently president of the ECC Alumni Association and received the "Jaycee of the Year" award in 1981. In 1985 he established a $500 annual scholarship in his mother's honor.

Julie A. Vanderheyden
Highland Community College, IL

Being one of the top students in my high school graduating class was an honor, but it also carried the burden of living up to expectations. Students, teachers, and community members would ask me where I planned to go to college and what my intended major would be. Of course they expected me to attend a major university and enroll in a course of study such as engineering.

When considering a career, I realized that my life would not be fulfilled unless I pursued my first love—the performing arts of music and theater. The more colleges and universities I visited, the more I realized that nearby Highland Community College would give me more opportunities to perform than would a large school. I could get valuable stage experience while freshmen at other colleges would dream of performing with the jazz ensemble or landing a role in a theater production someday. Highland's top-notch theater department and facility were comparable or superior to many of the four-year colleges that I visited.

When I announced Highland Community College as my college of intent, I surprised a lot of people. Some of my classmates teased me. One friend wrote in my yearbook, "Good luck at your next high school college!" But I knew that I would have the last laugh in the end when I graduated with my bachelor's degree in four years, debt free.

I double-majored in math and music education, but I also had the opportunity to take some theater courses and perform in many productions at HCC. I also performed with the jazz ensemble and pep band. My job at Highland as box office manager and musical accompanist not only gave me money to get through college, but also the skill and experience to get a job at Illinois State University.

The small class sizes and individual attention from caring instructors at Highland gave me an excellent educational foundation. I transferred to Illinois State University, where in two years I completed my bachelor's degree.

I am now in my seventh year as the instrumental music instructor for the East Dubuque Public Schools. I teach private piano lessons and direct a church choir. My theater experience at HCC has allowed me to direct the musicals for East Dubuque High School. I am in demand as an accompanist for local community and semi-professional theater productions.

Highland Community College gave me an excellent education while providing me with valuable performing experience. I made friendships and developed skills that will last a lifetime!

Julie A. Vanderheyden graduated from Highland Community College in 1980 with a degree in music/mathematics and now lives in East Dubuque, IL. She is a music instructor and continues to devote her talents to those in her community. She is constantly striving to keep abreast of the latest techniques in music education, as evidenced by her pursuit of a master's degree from Western Illinois University.

William Virchis
Southwestern College, CA

I was born in Mexico on Mother's Day, 1944. On that day when I saw the world for the first time, the bright lights of the operating room became a metaphor for the light I was to follow. I was born with my feet completely backwards. I took my first steps on my sixth birthday, after 72 operations on each foot.

I can't remember when I learned to speak English, but I can remember that special second-grade teacher who took time to tutor and take care of a new immigrant. Mrs. Attaway will always be with me: she was a great teacher.

At Chula Vista Junior High School another light went on...a light that was to influence my future. My eighth-grade homeroom teacher thought I had the talent and potential to be in show business.

At Chula Vista High School I was cast in a successful senior play. Then my drama teacher encouraged me to audition at the Old Globe Theater in San Diego. Thanks to a teacher another part of the light was bright. Thank you, Mrs. Lowe.

My art teacher, who was also my wrestling coach in high school, became a teacher the first year of Southwestern College's existence. He encouraged me to enroll at Southwestern College and talked me into staying in school. (I had been on my way to Los Angeles to become "a star"). He put it quite plainly: "Do you want to be an ignorant actor or a smart one? Stay in school." My Southwestern College history teacher, a Lebanese man with great vision and intellect, was also a great influence. He is now Sen. Wadie Deddah in the California State legislature.

I took a break from school—I worked in Los Angeles and traveled and worked with my father in LaPaz, Mexico. Then I came back to Southwestern College at the age of 25. I graduated from Southwestern College with my associate degree and then went to San Diego State University, where I majored in drama, Spanish, and psychology. The foundation of my community college experience paid off.

William Virchis graduated from Southwestern College in 1968 with a degree in theater arts and now lives in Chula Vista, CA. He is an instructor and director of theater arts at Southwestern and has directed numerous productions at his college and for the San Diego Civic Light Opera, Lyric Dinner Theater, San Diego Gilbert and Sullivan Repertory, California Pacific Community Theatre, Teatro Meta, and The Bowery Theatre. He staged the re-enactment of the Landing of Cabrillo for the U.S. Department of Interior and received an Atlas Award for his portrayal of Pancho in "The Night of the Iguana."

Later, I decided to go to New York to become an actor/director. But fate stepped in. Instead, I was offered a job at my old high school as a drama teacher, a real "Welcome Back Kotter" experience. I was then offered a job teaching at Southwestern College. The circle had been completed. When I stepped onto the stage in Mayan Hall at Southwestern College as a teacher, I thought back to the time when I was the first student on that stage.

Now, 16 years later, I believe that every student who passes through our doors creates the same energy that first turned on the light for me.

Patricia S. Wager
College of DuPage, IL

The fact that I am writing 20 years after graduation about the positive and exciting experience I had at College of DuPage as a student is tangible evidence that I am a believer in this institution. My transition from high school was an easy one. The registration process, my instructors, and the atmosphere were not intimidating to me. I was inspired by my teachers to excel. They were helpful, concerned, and always willing to go the extra mile to assist. This attention and quality of education translated into a better GPA at College of DuPage than in high school.

In addition to the outstanding academic experience, I was able to get involved in student organizations while I was a student at College of DuPage. For two years I actively participated on the student program board, planning social events and promoting student activities. I believe this experience helped to shape the direction of my career. I discovered that I enjoyed working with people and all of the details associated with program planning.

A year and a half after I graduated from College of DuPage, I had the opportunity to participate in organizing the College's first alumni association. I was pleased to give back in small measure what the College of DuPage had so generously provided to me while I was a student. The College of DuPage had touched my life in several important ways. Personally, I met my husband while attending College of DuPage; academically, I received an associate's degree and was given the base to continue my education and receive a bachelor's degree, cum laude; socially, I met and have maintained the friendship of many classmates and instructors; and professionally, I was given the inspiration to achieve my prescribed career goals. My 17 years of career experience have all been in higher education, and my positive experience at College of DuPage helped influence this career path.

I spent nine and a half years as a coordinator of alumni affairs at College of DuPage. I became director of development for the Loyola University School of Dentistry and eventually development officer for the Department of Ophthalmology at the University of Illinois at Chicago. Life-long learning is what the College of DuPage has meant to me.

The excitement continues as I watch my daughter study the College of DuPage class schedule and select the classes she will take during the summer of 1990.

> *Patricia S. Wager graduated from the College of DuPage in 1970 with a degree in general studies and now lives in Wheaton, IL. She is a development officer for the University of Illinois at Chicago's Department of Ophthalmology. She was instrumental in organizing and developing the college's alumni association in 1972 and has volunteered her time and special talents to her peers in numerous professional organizations.*

Carl S. Warren
Clinton Community College, IA

Since I was the second child in a family of five, my parents were not in a financial position to send me or my older brother to college at one of Iowa's four-year colleges or universities. Fortunately, I was awarded a scholarship to Clinton Community College. By working several part-time jobs during the academic year and working two full-time jobs during the summers, I was able to save enough to pay my way through my junior and senior years at the University of Iowa.

After completing my bachelor's degree, I was awarded a teaching assistantship at the University of Iowa, which allowed me to complete my master's degree. I was then awarded a graduate assistantship at Michigan State University and earned my doctorate. Without the opportunity that Clinton Community College provided me to continue my education directly out of high school, my college education would have been delayed, perhaps indefinitely. Community colleges provide a valuable service to society by giving students opportunities to continue their educations.

I have never regretted attending Clinton Community College. My first accounting instructor, Jack Whipple, provided me with a solid foundation in my first two accounting courses. After transferring to the University of Iowa in my junior year, I was surprised to find that I had a better foundation in the basics of accounting than many of their students. In fact, I often tutored their students in the basics. This solid foundation enabled me to earn "As" in all my accounting classes both at the undergraduate and graduate levels. It has been my pleasure to contribute to a scholarship in memory of Mr. Whipple at Clinton Community College.

Clinton Community College also nourished my quest for knowledge and built confidence in my abilities. I still remember the positive encouragement that my English instructor, Mel Erickson, provided during my freshman English classes. He made writing an easy and enjoyable experience. As a consequence, I have become co-author of the fourteenth edition of the introductory accounting textbook I initially used as a student at Clinton Community College. This textbook, *Accounting Principles* by South-Western Publishing Company, is the best-selling accounting textbook in the world. Becoming a co-author of the accounting text I used as a student is a satisfying experience that can be directly attributed to my educational training at Clinton Community College.

The small campus environment at Clinton Community College encouraged interaction among the students and faculty. This interaction enabled me to develop confidence in working with others, which helped me adapt socially throughout my career.

Carl S. Warren graduated from Clinton Community College in 1967 with a degree in accounting and now lives in Athens, GA. He is a professor of accounting at the University of Georgia. He is the author of more than 40 articles and 10 textbooks and has co-authored Accounting Principles, *a best-selling accounting textbook. As an accounting professor, he is passing on to hundreds of students the accounting principles he began to learn at Clinton Community College.*

James L. Wattenbarger
Palm Beach Community College, FL

Palm Beach Junior College, now Palm Beach Community College, provided me with opportunity beyond high school that I could not have had otherwise. The three barriers that interfere with education beyond high school graduation—the financial barrier, the geographical barrier, and the educational program barrier—were all very much present in my life at the age of 18. However, the low cost of community colleges, the ready accessibility of Palm Beach Junior College, and the solid academic program I studied there, were the factors that made a difference and enabled me to develop a professional career that is most rewarding, both in terms of an opportunity for service to the communities of Florida and in terms of my own personal development.

The sensitivity of those great faculty members at Palm Beach Junior College who provided me with the opportunity to prepare for the profession of education was a major factor in my following this goal. The local business and professional men who contributed to the Rotary Club scholarship that I received to assist me in continuing beyond community college graduation were also important influences. Finally, the financial support I received by driving a school bus (local assistance) and working as an NYA assistant to the Dean of Palm Beach High School (federal assistance) were most important factors.

The encouragement from the people in my own home community has been the major factor in the development of my career—one that has concentrated on securing similar support for others from people in their home communities, the faculty in the education profession, and various financial resources. An expression of thanks to those who made it possible for me to assist in the continuing development of community colleges is a major purpose of this statement—those at Palm Beach Community College as well as many friends in the American Association of Community and Junior Colleges.

James L. Wattenbarger graduated in 1941 from Palm Beach Community College and now lives in Gainesville, FL. He is a professor and director of the Institute of Higher Education at the University of Florida. His doctoral dissertation served as the blueprint for the Florida System of Community Colleges, which now includes 28 institutions. He is active in church affairs and has served as Boy Scout group committee chairman.

PROFILEES INDEX

COLLEGE INDEX

Texas

Vermont

Virginia

Washington

West Virginia

Wisconsin

About the Editors

Robert E. Bahruth received his AA degree in 1970 from Middlesex County College, NJ, and his teaching degree from Tusculum College, TN, in 1972. He completed an MA in bilingual education and English as a Second Language at the University of Texas at San Antonio. At the end of that program he began his doctoral studies in curriculum and instruction, minoring in linguistics at the University of Texas at Austin. He has taught high school English, and he has taught ESL stateside and overseas in Latin America. He also taught in a fifth grade bilingual classroom in Pearsall, TX. While in Austin, he taught ESL with Phil Venditti at Austin Community College. They began sharing ideas, writing articles, and presenting at professional conferences together. Bahruth is an assistant professor of bilingual education at Boise State University, ID.

Phillip N. Venditti received a BA degree in 1971 from the University of Colorado. Three years later, after studying and working in Norway, Germany, and the Netherlands, he began graduate study at the University of Tennessee, which culminated in an MS in English education. As a Peace Corps volunteer from 1976 to 1978, he worked as an English teacher, teacher-trainer, and researcher in the Republic of Korea. After fulfilling requirements for a master's in international administration from the School for International Training, he worked five years as a student services administrator in four-year colleges and universities. He joined "The Block of 1985" in the Community College Leadership Program at the University of Texas at Austin and completed his doctorate there while teaching ESL at Austin Community College. He has worked as a community college teacher and administrator ever since and is currently associate dean for humanities and social sciences at Genesee Community College, NY.